SURFACE IRRIGATION
systems and practice

SURFACE IRRIGATION
systems and practice

Melvyn Kay
BSc, MSc, CEng, MICE

Cranfield Press
Cranfield Institute of Technology
Cranfield Bedford MK43 0AL England UK

First published 1986

Surface Irrigation:
Systems and Practice

Cranfield Press
Cranfield Information Technology Ltd
Cranfield Bookshop
College Road
Wharley End
CRANFIELD
Bedford MK43 0ST
UK

ISBN 0 947767 26 6

Printed in Great Britain by Cotswold Press, Oxford.

CONTENTS

CONTENTS continued

FOREWORD

Although 1,000 million hectares of land in the world is cultivated, only 20% of this cultivated area is irrigated. Yet this irrigated area produces more than 40% of the world's agricultural output. Thus the ability to produce more food will depend, to a large extent, on extending the area under irrigation and improving existing systems.

In recent years, there has been a rapid increase in the use of sprinkler and trickle irrigation. However, the most popular irrigation method is still traditional surface flooding, whereby water is spread across the land using basin, border and furrow techniques. More than 90% of all irrigation involves the use of surface flooding and it is unlikely that this will change in the foreseeable future.

In this new book, Melvyn Kay recognises that for irrigation to be effective it not only has to be well designed, but also well managed.

The extensive programme of teaching and research in the field of irrigation at Silsoe College is directed towards achieving more effective and efficient systems in practice.

It is very much part of the whole ethos of Cranfield that this book should, therefore, be supported by Armfield Technical Education Company Limited, a leading irrigation company. We are grateful to them for their cooperation.

Above all, I would like to congratulate my colleague Melvyn Kay for producing a book which will make an important contribution to alleviating one of the key problems facing world agriculture.

Professor Brian May
Head of Silsoe College
Cranfield Institute of Technology

April 1986

PREFACE

This is the second book I have written primarily for those involved in irrigation at vocational level in the developing countries. It deals with surface irrigation which is the most widely used method of irrigation and is likely to remain so in the foreseeable future. Practical information is provided on methods of applying water, land preparation, open channels, structures, and pipelines for both large and small-scale irrigation schemes. I have also included a chapter on hydraulics which is intended for those with a limited technical or mathematical background; it describes not only the way water flows in channels and how it can be measured, but also how water moves on and in the soil during irrigation. The latter I feel is of great importance for anyone wishing to acquire even the most basic practical skills in surface irrigation.

Many people still believe that a good physical irrigation system means good irrigation practice, but this is clearly not so. For this reason I have tried to describe many of the common problems of irrigation practice and indicate ways in which they might be overcome. I have also devoted two chapters to the subjects of operation and maintenance of irrigation systems. The inability to carry out these tasks properly is at the root of many of the problems we see on irrigation schemes in the developing countries.

Lastly I have included a chapter on health problems in irrigation. Not enough attention is given to this and as more and more schemes are constructed, diseases which are related to water also spread. Those in irrigation must be made aware of these dangers and the ways in which they can contribute to bringing these diseases under control. I hope that in some small way this chapter will contribute to increasing that awareness.

Although I have aimed the book at the practical man in the field, I hope that many aspects of the text will be of interest to engineers, agriculturalists and others involved in irrigation at both middle and professional levels.

As in my previous book I have used diagrams and photographs wherever possible to try and overcome some of the difficulties that many students have in studying English texts when this is not their first language.

M G Kay
Silsoe 1986

ACKNOWLEDGMENT

Special thanks are due to Armfield Technical Education Co Ltd for their support in publishing this book; to George Dilks for preparing the artwork and to my wife, Judith for her encouragement during the preparation of the manuscript and for preparing the index.

The author and publisher would also like to thank the following individuals and firms for supply photographs: M K V Carr for figures 1.3(c), 3.3(a), 5.9(a); B Cornut for figure 5.5; Food & Agriculture Organisation for figures 1.1(b), 2.7(b), 3.3(b), 4.4(c), 5.7(c), 5.9(b), 9.11(b), 10.2(c) and (d); N W Hudson for figure 2.4(c), J Jewsbury for figures 13.2, 13.3(a) – (c), 13.5(a) and (b), 13.6, 13.7; H Leverton & Co Ltd (Caterpillar) for figures 7.7, 7.8, 7.10(b); J H Neville for figures 3.3(c), 9.10(b), 9.14(a); Overseas Development Unit, Hydraulics Research Ltd for figures 1.2(c), 7.6(b) and (c), 12.4; Ruston-Bucyrus Ltd for figures 12.1(a), 12.2(c); Spectra-Physics Ltd for figure 7.11(b); G Spoor for figure 1.1(a) and Waterman Industries Inc. for figures 1.1(c), 9.7(b), 10.1(b) and (c), 10.2(b), 10.4(a) – (e).

1

INTRODUCTION

Irrigation is practised in many parts of the world to provide water for crop growth. In dry areas such as the Middle East, India, Western USA and parts of Australia there is little or no rainfall and irrigation must provide the total water needs of the crop. In humid and temperate areas such as Central Africa, South East Asia and Europe, although crops can be grown under natural rainfall it is often insufficient and poorly distributed. Extra water is sometimes applied, supplementing rainfall, to improve crop yields and quality.

Of the total land area of 1000 mha being cultivated in the world only 200 mha are irrigated (20%). This produces more than 40% of the world's agricultural output. The need to produce more food will thus depend to a large extent on enlarging the area under irrigation and improving existing schemes.

1.1 *Irrigation methods. (a) Basin irrigation*

Although in recent years there has been a rapid increase in the use of sprinkler and trickle irrigation, the traditional method of surface flooding is still the most widely used. Water is spread across the land surface using basin, border and furrow methods (figure 1.1). More than 95% of all irrigation is done in this way and is likely to remain so for the foreseable future.

Many types of surface irrigation schemes have been developed to suit different agricultural systems and communities. Some are very small, covering only a few hectares, others are large, covering many thousands of hectares. However, they all have the following components in common:

Irrigation methods

Irrigation units

Irrigation system

Drainage system

Access

1.1 Irrigation methods

These are the methods used to spread water as uniformly as possible in the soil to irrigate the crops. Although there are many ways of doing this there are three main methods (figure 1.1):

Basin irrigation (chapter 3)

Border irrigation (chapter 4)

Furrow irrigation (chapter 5)

1.1(b) Border irrigation

1.1(c) Furrow irrigation

Many factors affect the choice of irrigation method such as land topography, soils, field shape, crops and labour (see chapter 6 *Choosing an irrigation method).*

1.2 Irrigation units

These are areas of land on which crops are irrigated by the various irrigation methods.

On small irrigation schemes a unit may be only one field or an entire farm (figure 1.2).

On large schemes, a unit may be one large field (eg sugar cane estate) or several small farms grouped together (eg small holder development) (figure 1.3).

The size of irrigation units can differ greatly. On some schemes units may be only a few hectares. On the larger schemes they can be 70 ha or more.

Water is distributed within each unit in small channels and small hydraulic structures are used to control and measure the flow to the fields.

On some schemes pipes are now used instead of channels. These may be permanently installed above or below ground and made of concrete, asbestos-cement or plastic, or portable pipes made of aluminium or plastic which can be moved from one part of a unit to another (see chapter 10 *Pipelines*).

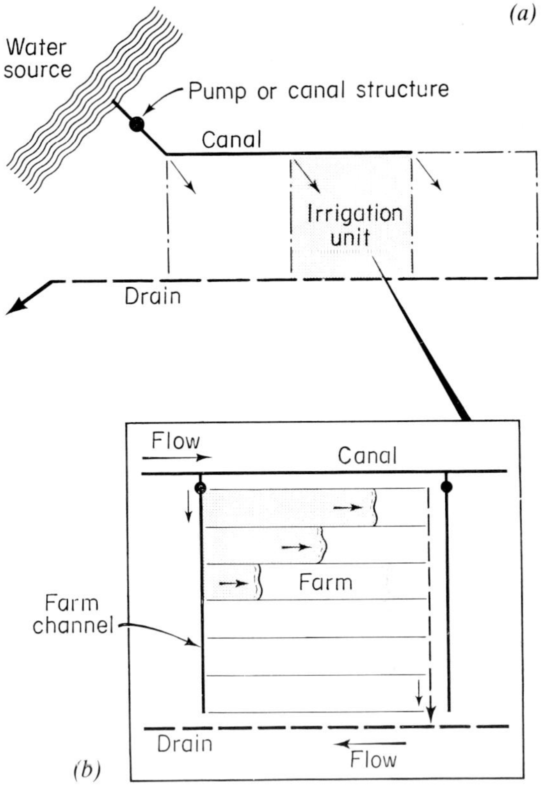

1.2 Small irrigation scheme. (a) General plan (b) Irrigation unit (c) Typical small scale rice scheme

(c)

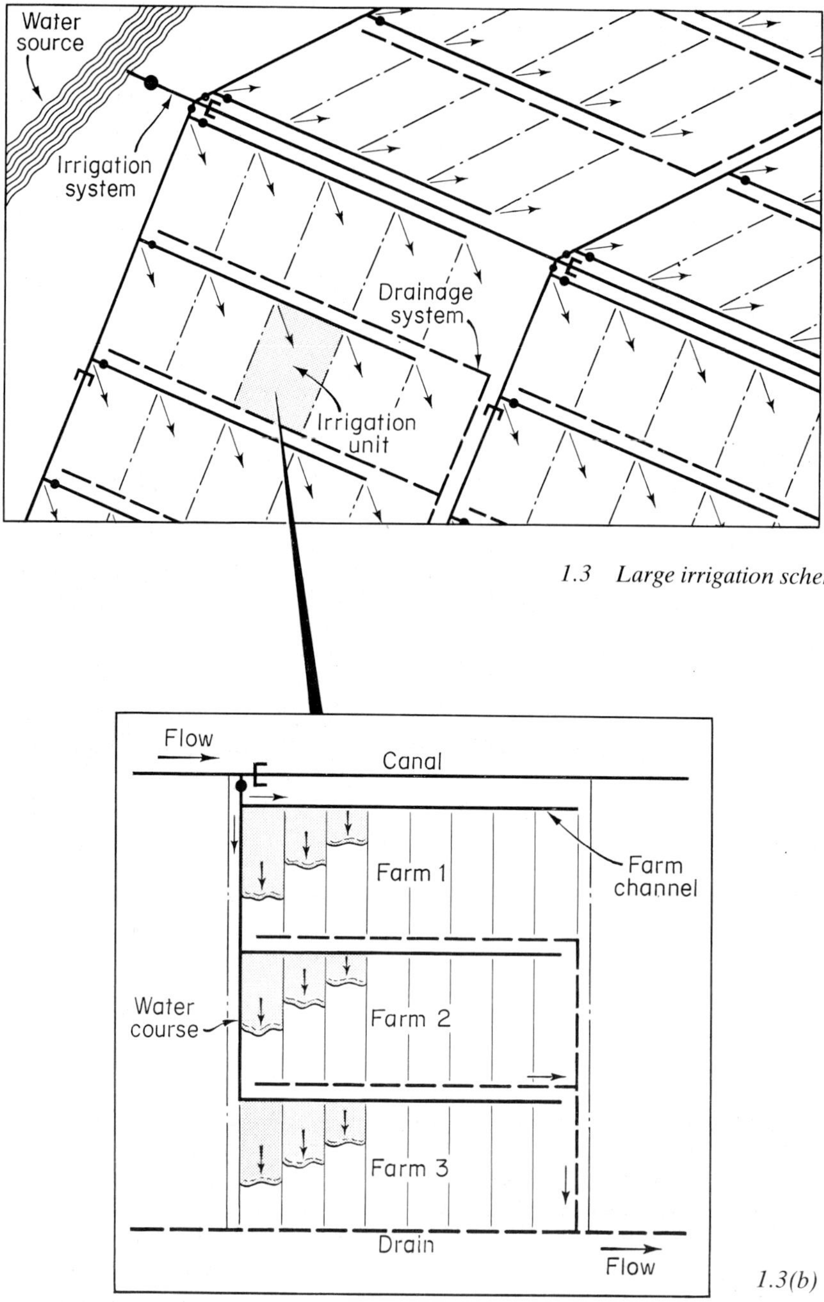

1.3 Large irrigation scheme. (a) Part plan

1.3(b) Irrigation unit

1.3 Irrigation system
This is usually a system of channels, called canals, used to distribute water from some source, such as a river or borehole, to the irrigation units.

On small schemes a single canal may be sufficient (figure 1.2). On large schemes a more complex system of canals is needed (figure 1.3).

Canals are usually constructed by forming embankments from the surrounding soil. In areas of sandy soils, seepage losses can be high and so canals are often lined with concrete, bitumen or plastic sheeting (see chapter 8 *Open channels*).

Hydraulic structures are used in the canals to measure and control the distribution of water so that each irrigation unit receives its

1.3(c) Aerial view

share of the flow (see chapter 9 *Hydraulic structures*).

Although most irrigation schemes use canals, some use pipelines for part or all of the irrigation system. They are usually permanently installed underground and made of galvinised steel, asbestos-cement, concrete or plastic (see chapter 10 *Pipelines*).

1.4 Drainage system

No irrigation can be complete without a drainage system (figure 1.2 and 1.3).

Unwanted water can accumulate on fields from heavy rainfall or excessive irrigation. Too much water applied in this way can be very harmful to crops and so must be removed as quickly as possible. This is usually done by a system of pipes or open channels which drain the water from the fields. Large open channels take the surplus water away from the irrigated area for disposal.

On some irrigation schemes the drainage water is re-used for irrigation.

1.5 Access

Good roads are an important part of any irrigation scheme and should be included in the planning of the irrigation and drainage system. Farmers need access to their fields for farm machinery and lorry transport may be required to take produce from the farm to the markets.

Major roads in a scheme may be paved. Minor roads and access to individual farms may be just graded tracks. Embankments formed when constructing canals and drains usually make good roads and keep vehicles out of wet, low-lying areas.

1.6 Operation and maintenance

When an irrigation scheme has been constructed it must be operated properly for it to be successful. This means delivering the right amounts of water to the farms at the right time. On small schemes this can be a simple task but on large schemes there are many problems in organising and distributing water (see chapter 11 *Operation)*.

Maintenance too is important for the irrigation and drainage systems to work properly. This is the routine work which must be done such as cleaning and repairing canals, drains, structures and cleaning and lubricating gates (see chapter 12 *Maintenance)*.

1.7 Health

Not enough attention is being given to this problem in irrigation. As more and more schemes are constructed the diseases which are related to water are also spread. Irrigators must be aware of these problems and the ways in which they can reduce the risk of infection (see chapter 13 *Health and Irrigation)*.

The following chapters describe in more detail some of these component parts of surface irrigation schemes. Details are given about their advantages and disadvantages and how they can be used and maintained to provide efficient and useful service.

2

HYDRAULICS

2.1 Objectives

Irrigation water is supplied from a source (eg well, river or reservoir) through canals or pipelines and spread as uniformly as possible over the land. The detailed design of a scheme is usually carried out by an experienced engineer working closely with agriculturalists and soil scientists. His job is to plan the layout of the irrigation system, select the right sizes for canals and pipes, select the structures to use to control water flow and select the method of irrigation. To operate an irrigation scheme it is not essential to understand the complexity of its design. However, a working knowledge of how water flows in canals and pipes[1] and how it spreads across the land and infiltrates into the soil will help the irrigator to make full and proper use of his scheme.

[1] Flow in pipes is described in a companion volume *Sprinkler Irrigation: equipment and practice*

2.2 SI Units (international metric system)

SI units are slowly replacing Imperial and European units in irrigation but the latter systems are likely to be around for some time yet. Indeed a good deal of confusion exists as irrigators tend to use more than one system of units. For example, pipe sizes may be given in SI units, eg 75 mm or 100 mm diameter, and yet pump capacities may be given in Imperial units, eg 300 gallons per minute. To avoid such confusion in this text the SI system is used throughout.

The fundamental units of the SI system are:

Measurement	*Unit*	*Symbol*
Length	metre	m
Area	square metre	m^2
Volume	cubic metre	m^3
Mass	kilogramme	kg
Force	newton	N

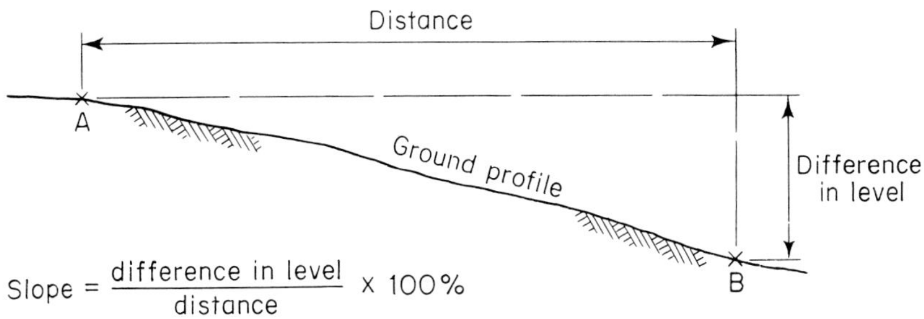

2.1 Measuring land slope.

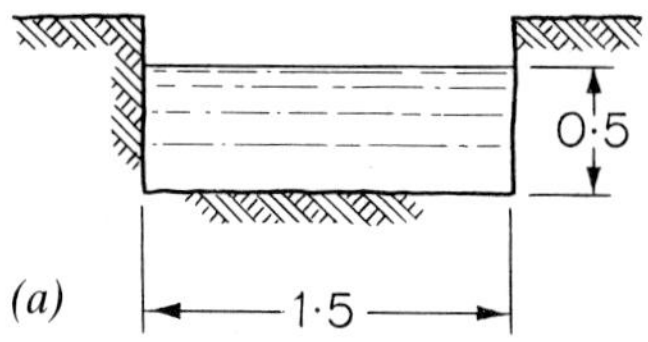

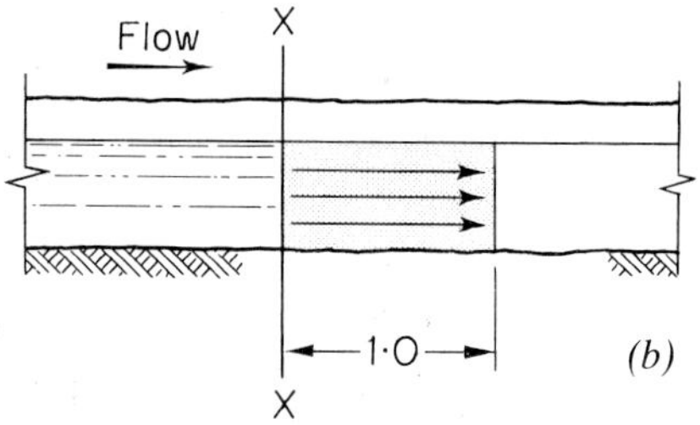

2.2 Calculating discharge in a channel. (a) Cross section (b) Longitudinal section

2.3 Land slope

This is a measure of the difference in ground level between two places in a field and is referred to as a percentage (%) (figure 2.1). It is the number of metres difference in elevation for each 100 m of horizontal distance.

As water will only flow downhill it is important to know which way the land is sloping when setting out new channels and field irrigation methods. This may seem obvious but in areas where surface irrigation is used the land slopes are usually very small (0-2 %) and are not easily detected by eye. Special survey instruments are needed to determine slope. It is all too easy to set out border and furrows in the wrong direction and end up trying to make water flow uphill. An example of this kind of problem is shown in figure 7.12.

FLOW IN CHANNELS

2.4 Discharge

The speed at which water flows in a channel is called the *velocity*. It is measured in metres per second (m/s). The *discharge* is the volume of water flowing each second and is measured in cubic metres per second (m^3/s). To understand this, consider the case of water flowing in a canal 1.5 m wide and 0.5 m deep (cross sectional area 0.75 m^2) at 1.0 m/s (figure 2.2). In one second the water flowing past x - x will be the shaded volume. The length of the shaded portion is numerically equal to the velocity – in this case 1.0 m. The volume flowing each second (the discharge) is equal to the canal area multiplied by the shaded length, ie $0.75 \times 1.0 = 0.75$ m^3/s.

In more general terms:

Discharge (m^3/s) = cross-sectional area (m^2) × velocity of flow (m/s)

When the discharge is small, as in farm channels, it is often measured in litres per second (ℓ/s). In the example given 0.75 m^3/s is equal to 750 ℓ/s. The conversion is made by multiplying by 1000.

The discharge required to irrigate basins, borders and furrows is usually called the *stream size*. This is measured in litres per

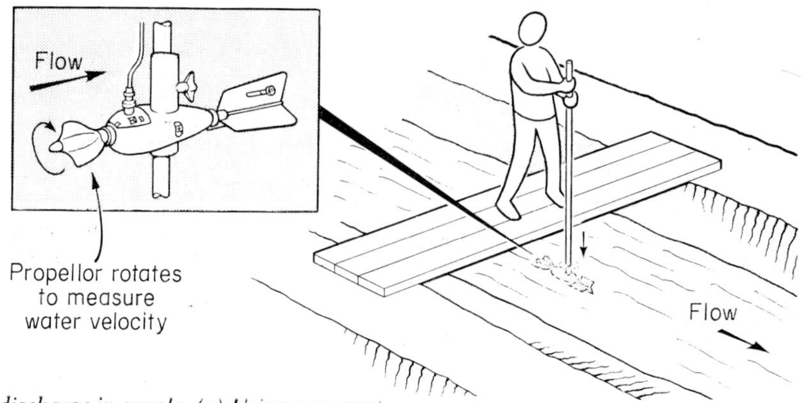

2.3(a)

2.3 Measuring discharge in canals. (a) Using a current meter (b) Using the float method (c) Using the volume method

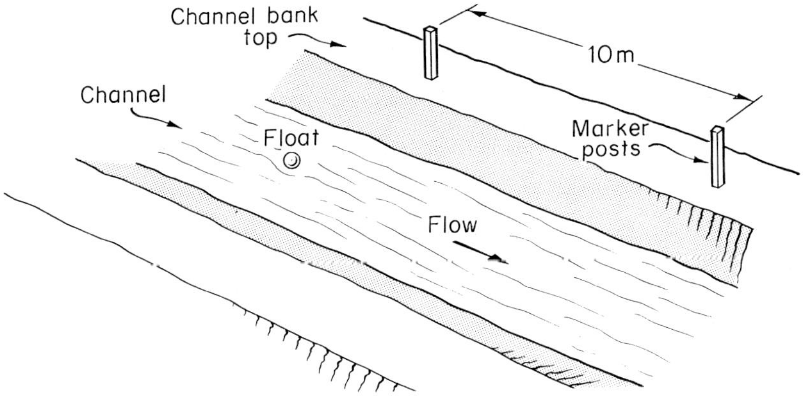

2.3(b)

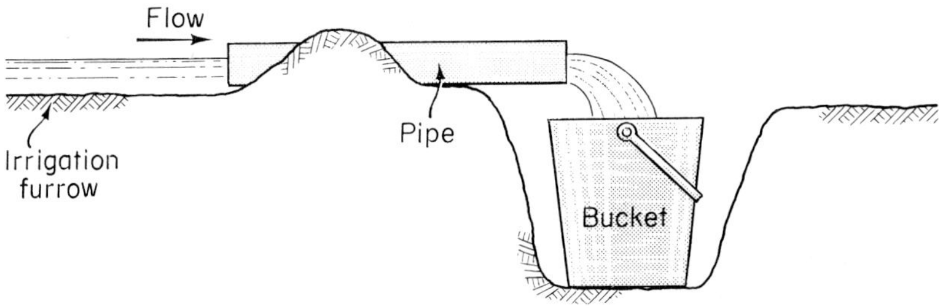

2.3(c)

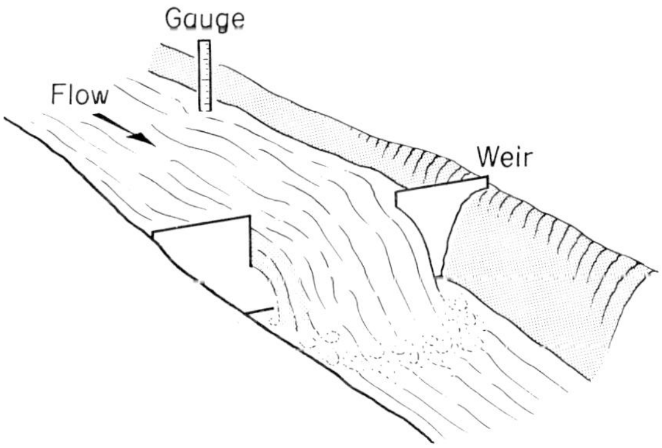

2.4(a)

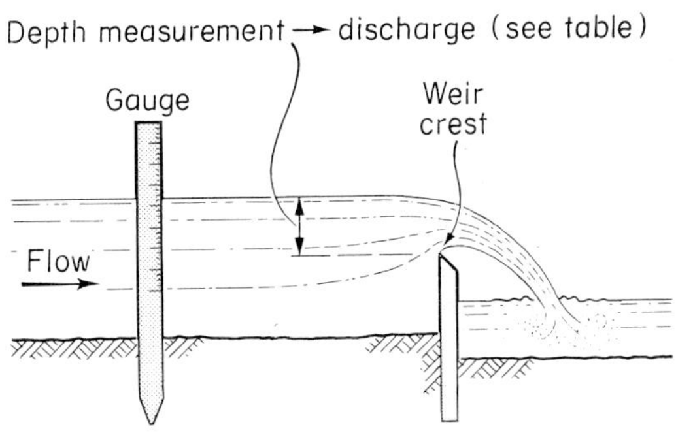

Head (mm)	Discharge* (L/s)
30	9·5
40	14·6
50	20·4
60	26·7
70	33·6
80	40·9
90	48·9
100	57·0
150	103·8
200	158·5

* This is for a weir crest length of 1·0m

2.4(b)

2.4 Measuring discharge using weirs and flumes. (a) Rectangular weir (b) Rectangular weir details

2.4(c) Parshall flume installed in a main canal

second. In borders the term *unit stream* is used. This is the discharge for each metre width of the border. The stream size is equal to the unit stream multiplied by border width.

2.5 Discharge measurement

Current meter method
Current meters are instruments for measuring water velocity (figure 2.3). The meter is lowered into the channel and the flow causes the propellor to rotate . The speed of rotation is proportional to the velocity of the water and so it can be measured by recording the number of turns of the propellor in a given time. Several measurements are needed across the channel to determine the average water velocity. The discharge is calculated by measuring the cross sectional area of the channel and using the above formula.

This method is very suitable for large canals.

Float method
This is a simple but approximate way to measure discharge. Two points are marked along the bank of a straight length of channel 10 m apart (figure 2.3). A float such as a rubber ball or a stick of wood is then thrown into the middle of the channel and the time taken for it to travel the 10 m is measured in seconds. The velocity is calculated by dividing the distance (in this case 10 m) by the time. Remember that the float must be in the middle of the canal. This is because water always flows much slower near the canal banks due to friction. The measurement should be repeated several times to avoid any errors. The discharge is calculated by measuring the cross sectional area of the stream and using the above formula.

Volume method
This method is used to measure discharge in very small channels such as an irrigation furrow (figure 2.3). A hole large enough to

take a 5 litre bucket is dug in the furrow. A small length of pipe is buried in the furrow so that all the water passes through it. The pipe needs to be large enough so as not to restrict the flow and to stop the furrow being overtopped. The time taken to fill the bucket is noted and the volume divided by the time is equal to the discharge.

$$\text{Discharge } (\ell/\text{s}) = \frac{\text{volume of bucket } (\ell)}{\text{time taken to fill (s)}}$$

Weirs and flumes
These are structures built temporarily or permanently in a channel to measure discharge. They are more accurate than the previous methods described.

There are many different types of weir. One of the most common is the rectangular weir (figure 2.4). This consists of a sheet of wood or metal in which a rectangular notch has been cut. The weir is placed across the channel so that all the water passes over the notch or weir crest. A gauge is placed in the channel upstream of the weir so that the depth of water in the channel above the weir crest can be measured. Using a table of known values this depth is used to determine the channel discharge.

Weirs of this type are very useful for installing temporarily in farm channels to measure how much water is being used on farms (figure 11.6). For more long term measurements a permanent structure is needed as small weirs of this kind are easily damaged.

There are also many different types of flumes. These are specially constructed narrow sections of channel. A typical flume used in irrigation is the Parshall flume (figure 2.4). The discharge is determined by observing the depth of water in the channel upstream of the flume and using a table of known discharge values.

The reader should refer to texts on hydraulic structures for more detailed information on weirs and flumes.

2.6 Factors affecting flow

There are three factors which affect the way water flows in channels. These are:

Slope

Roughness

Shape

Slope
Water will only flow downhill in an open channel and the steeper the slope the faster the water will flow. If the slope becomes very steep erosion can occur.

Roughness
The bed and sides of a channel can be quite rough and this slows down the flow in the same way as friction slows down an object as it is pushed across some rough surface (figure 2.5). A canal lined with smooth concrete has much less friction than an earth canal. This means that a lined canal can be much smaller than an earth canal and still carry the same discharge.

Weeds growing in a badly maintained channel will greatly increase roughness and can stop the flow completely (figure 12.3).

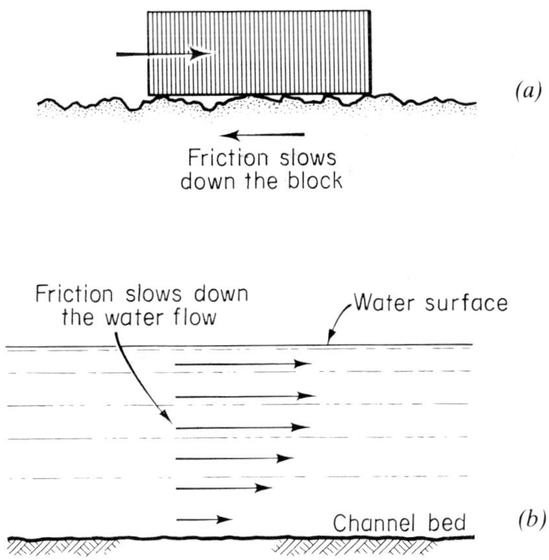

2.5 Effects of roughness on channel flow. (a) Block on rough surface (b) Flow in a channel

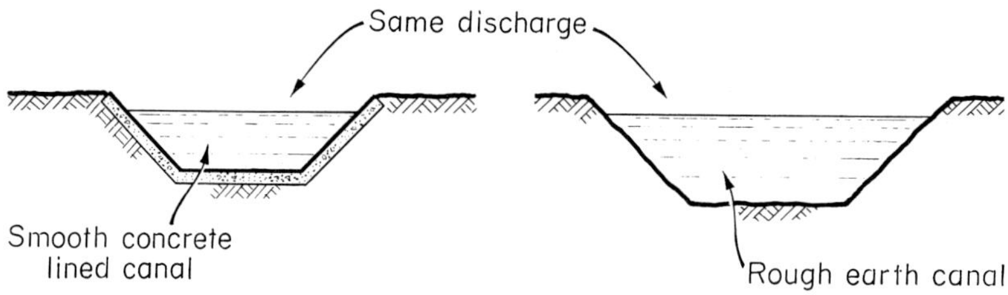

(c) Roughness affects canal size

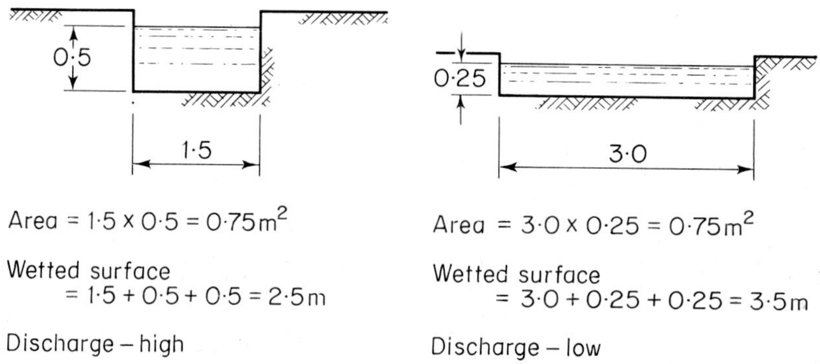

2.6 *Effect of channel shape on flow.*

Shape

Channels can have the same slope, roughness and cross-sectional area but carry different flows. This is because the channels can have different shapes. Consider the case of water flowing in two channels with the same area of $0.75m^2$ (figure 2.6). One channel is wide and shallow and has a surface length of 3.25 m in contact with the flow. The other channel is narrow and deep and has only 2.0 m in contact with the flow. There will be more friction in the wide channel than in the narrow one and so the water velocity will be less. This means the discharge will also be less.

This effect is sometimes used when designing channels on steep slopes where soil erosion can be a problem. Channel shapes are made wide and shallow to slow down the flow and stop any scouring of the sides and bed.

Selecting channel size

The relationships between flow, slope, roughness and shape are complex and it is the design engineer's task to understand and use them to select the right size of channel to use.

2.7 Siphons

Siphons are lengths of pipe used to supply water from a farm channel to basins, borders or furrows (figure 2.7). They are made from plastic or aluminium and are sometimes shaped to fit over the bank of the farm channel.

To use a siphon it must first be filled with water to take out all the air and then laid over the channel bank. This must be done quickly and a hand placed over the end of the pipe will stop air from re-entering. Provided the water level in the channel is above that in the field then water will flow out. Starting siphons in this way is know as priming. Irrigators with long experience make siphon priming look very easy but it takes lots of practice to develop the skill properly without getting wet.

Siphon discharge depends on the size of the pipe and the *head* on the siphon – this is the difference between the channel and field water level (mm). As the head increases the discharge through the siphon also increases. Typical siphon sizes range from 10 to 150 mm in diameter. Discharges for the smaller more commonly used siphons are given in table 2.1.

Siphons are an alternative to the untidy but common practice of breaking open the farm channel bank with a spade to release water.

(a)

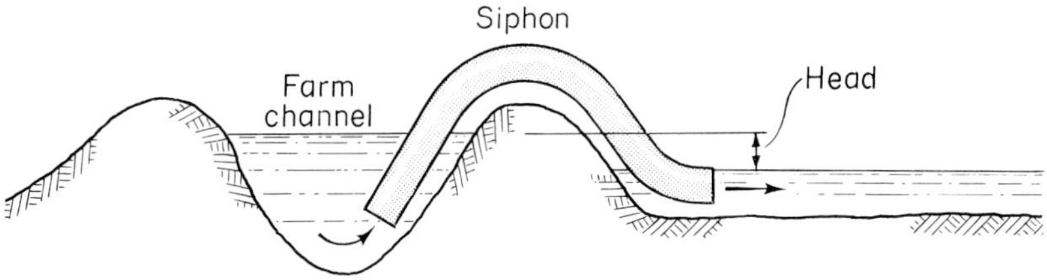

(b)

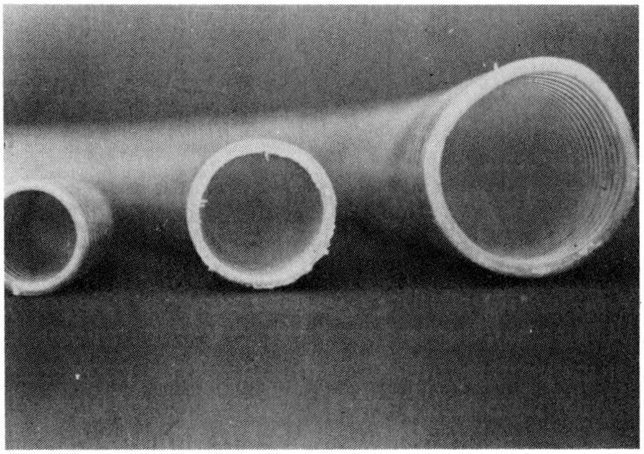

(c)

2.7 *Siphons. (a) Operating conditions (b) Different pipe sizes (c) Transporting siphons (d) Measuring discharge*

They are inexpensive and portable and so an irrigator may only require a few to irrigate a large area. The discharge into a basin or furrow can be easily controlled by using one or more siphons to make up the required flow.

Example

A farm channel carries a flow of 20 ℓ/s for a field using furrow irrigation. If each furrow requires a flow of 1.0 ℓ/s how many furrows can be irrigated at a time? If the head available is 100 mm what size of siphon is required?

$$\text{Number of siphons required} = \frac{\text{total flow}}{\text{furrow flow}}$$

$$= \frac{20}{1.0}$$

$$= 20 \text{ siphons}$$

From table 2.1, the best siphon size is 40 mm. Note that at 100 mm head each siphon will supply 1.06 ℓ/s. When all twenty siphons are working the flow would be 20 × 1.06 = 21.2 ℓ/s. This is greater than the flow available. In practice the channel water level would fall slightly and the flow in each siphon would fall to 1.0 ℓ/s as required. The outflow from the channel cannot be greater than the inflow.

Table 2.1 Discharges for siphons and spiles (ℓ/s)

Size (mm)	*Head* (mm) 50	100	150	200
10	0.05	0.07	0.08	0.09
20	0.19	0.26	0.32	0.73
30	0.42	0.59	0.73	0.84
40	0.75	1.06	1.29	1.49
50	1.17	1.65	2.02	2.33
100	4.67	6.60	8.09	9.34

Flow measurement

It is difficult to measure the head in the field and so an easier way to measure the discharge from a siphon is to use the volume method. However, the water must not flow directly into the bucket. Instead, it must first run into the hole and then spill over into the bucket (figure 2.7). The hole is first baled out and the bucket held firmly in place with its lip close to the normal water level in the field. As the water rises in the hole it starts to spill over into the bucket and the time taken to fill it is noted. The siphon discharge is found by dividing the volume in the bucket by the time taken to fill it.

Note that if the siphon flowed directly into the bucket the head and hence the discharge would be continuously changing as the bucket filled. The result would not be a true measure of the siphon discharge in its normal operating condition.

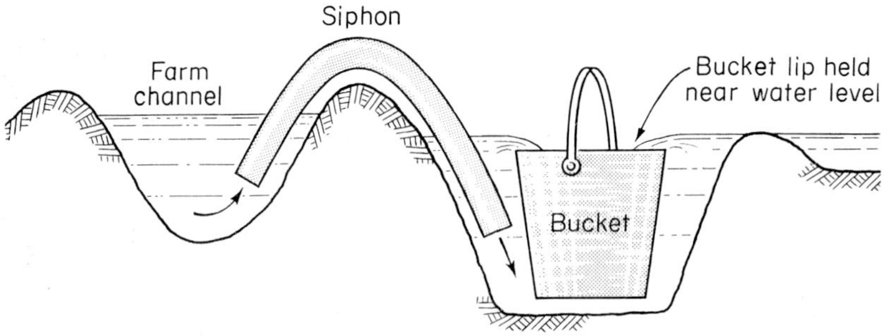

2.7(d)

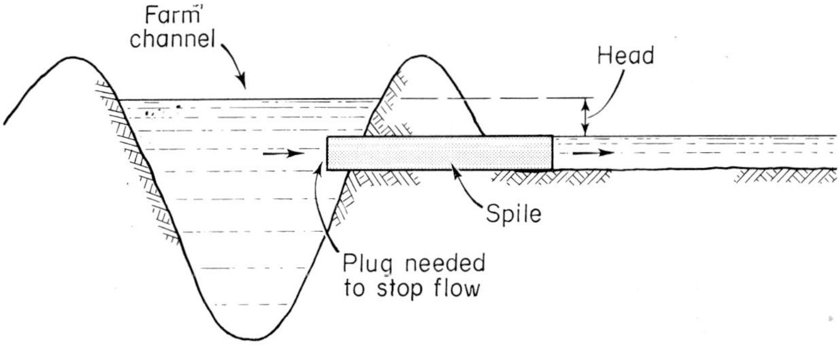

2.8 *Spiles.*

2.8 Spiles

These are also used to supply water from a farm channel into the field. They are short lengths of pipe buried in the channel bank (figure 2.8). They are made from rigid plastic or metal but sometimes local materials such as hollowed wood or bamboo are used. A plug is needed to close the pipe when it is not in use.

The discharge depends on the pipe diameter and the head of water available. Sizes and discharges are similar to those of siphons in table 2.1.

The discharge can be measured in the field using the volume method.

SURFACE IRRIGATION

Good surface irrigation practice involves:

Distributing water evenly in the soil

Providing adequate water for the crops

Avoiding undue water wastage and long term problems such as soil erosion and salinity.

The best crops are grown by providing an adequate supply of water and spreading this as uniformly as possible over the land. If not enough water is available then crops will suffer and yields may be reduced. Too much water can also reduce yields by damaging the crops and soils and causing waterlogging. Water is usually a limited and costly resource and should be used with great care to avoid wastage.

2.9 Soil reservoir

A field can be thought of as a reservoir or storage tank (figure 2.9). It is first filled with water by the irrigator and then slowly emptied by the plants. Just like any reservoir it can only hold a certain amount of water and no more. If too much water is poured in it overflows and is lost. In the soil, water can be lost in two ways. It can run off the surface into a drain - *run-off,* or it can percolate down into the soil beyond the roots of the crop - *deep percolation.* Ways of avoiding these losses must be found if water is to be used efficiently.

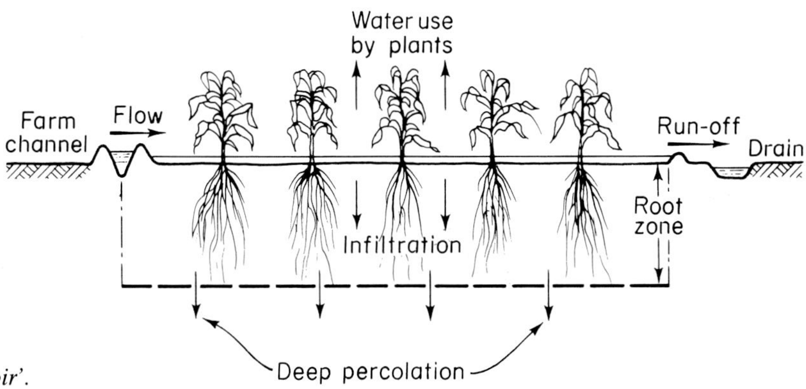

2.9 *Soil 'reservoir'.*

(a)

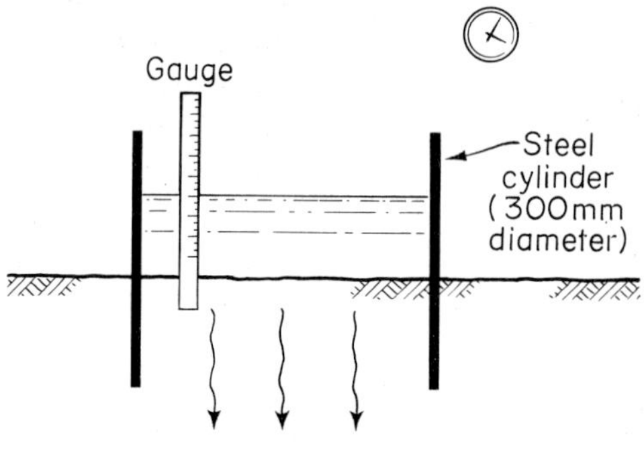

(b)

(c)

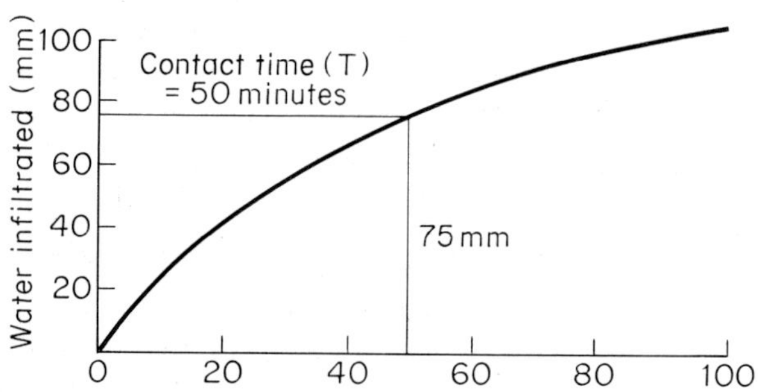

Water (mm)	Time (mins.)
15	5
25	10
40	20
52	30
74	55
90	70
103	100

Typical results

2.10 Measuring infiltration. (a) infiltration test (b) Field measurement (c) Typical results

2.10 Infiltration

The soil reservoir is filled by flooding water onto the soil surface so that it can be absorbed by the soil. Absorbing water in this way is called *infiltration*. Initially when the soil is dry, water infiltrates quickly. After 20-30 minutes this decreases as the air spaces in the soil become filled with water. After a period of 1-2 hours water infiltrates at a slow steady rate. This is called the *basic infiltration rate*. It is used to describe different soil types from an irrigation point of view because infiltration rate depends on soil texture (size of soil particles) and soil structure (the way in which soil particles are grouped together (table 2.2). The larger the soil particles and the further they are apart, the higher the infiltration rate. Other factors which can affect infiltration are tillage conditions, crop cover and land slope.

Infiltration rate is very important when deciding which method of irrigation to use. Soils with a low (0-10 mm/h) or medium (10-30 mm/h) infiltration rate are suitable for surface irrigation. Those with a high infiltration rate (greater than 30 mm/h) may only be suitable for sprinkler or trickle irrigation. On such soils, water is absorbed too quickly and it becomes difficult to apply water uniformly and efficiently with surface methods.

Table 2.2 Basic infiltration rates for soils

Soil type	*Basic infiltration rate* (mm/h)
Sand	30
Sand loam	20 – 30
Silt loam	10 – 20
Clay loam	5 – 10
Clay	1 – 5

Measuring infiltration

Infiltration can be measured by a simple field test. A steel cylinder 300 mm in diameter is forced into the soil and filled with water (figure 2.10). A gauge is placed in the cylinder to measure the fall in water level in millimetres as water infiltrates into the soil. The time in minutes is also recorded as each water level measurement is taken.

Typical results for a sandy soil are shown in figure 2.10.

Contact time

This is the time needed for water to infiltrate into a soil. It depends on the depth of water and the soil type. The results of the infiltration test can be used to determine the contact time. As an example, suppose 75 mm of water is to be applied to a sandy soil. Using the results in figure 2.10 the contact time would be 50 minutes. On a clay soil the contact time would be much longer because the infiltration would be slower.

When irrigating, the contact time must be sufficient to fill the soil reservoir. If it is too long then too much water will be applied and some will be lost. If it is too short the soil will be under-irrigated.

The contact time must be the same over the whole field. If it is different in one part of the field than in another, the application of water will be uneven.

2.11 Water movement in the soil

Water is normally supplied to the field through siphons or small outlets placed in the bank of the farm channel. It flows away from the channel and floods over the soil surface (figure 2.11 and 4.1). The water moves across the field and is confined by earth bunds on either side. This part of the irrigation is called the *advance*.

If it was possible to look below the soil surface water would be seen infiltrating into the soil reservoir (figure 2.11). Clearly the irrigation is not very uniform. Near the farm channel water has been infiltrating for some time and the soil reservoir is nearly full.

(a)

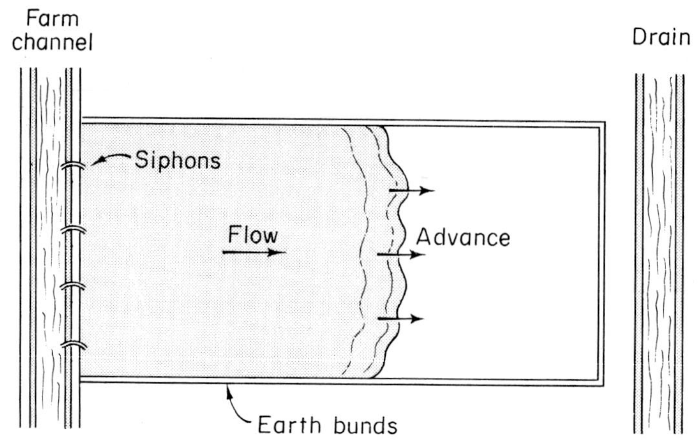

(b)

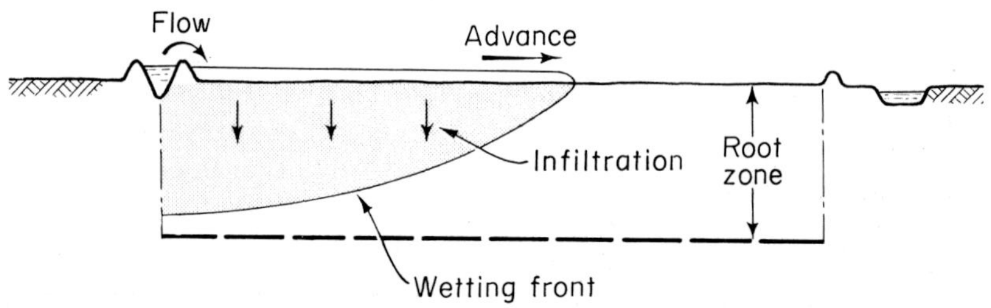

2.11 Water movement in the soil. (a) Water moving on soil surface (b) Water moving in soil profile

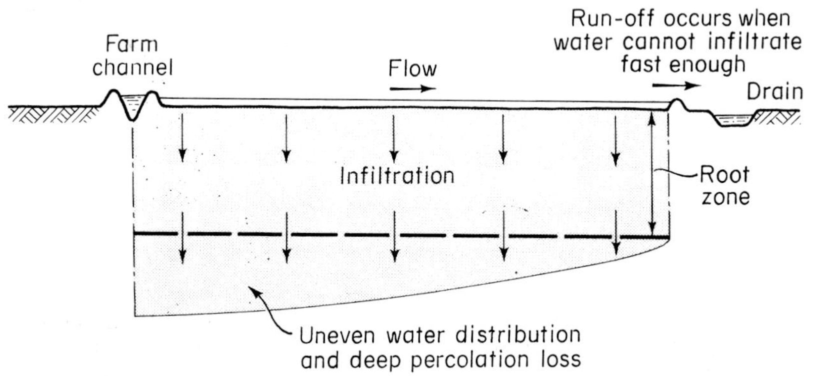

2.12 Surface irrigation problems.

Further along the field infiltration has only just started. To obtain a more uniform irrigation the flow must continue until water reaches the far end of the field and has sufficient time to infiltrate. However, this can cause two problems. Firstly, the contact time at the top of the field will be longer than at the far end and this will result in uneven water distribution and deep percolation (figure 2.12). Secondly if the water reaches the far end of the field and cannot infiltrate fast enough surface run-off can occur. Ways of overcoming these problems depends on the method of irrigation.

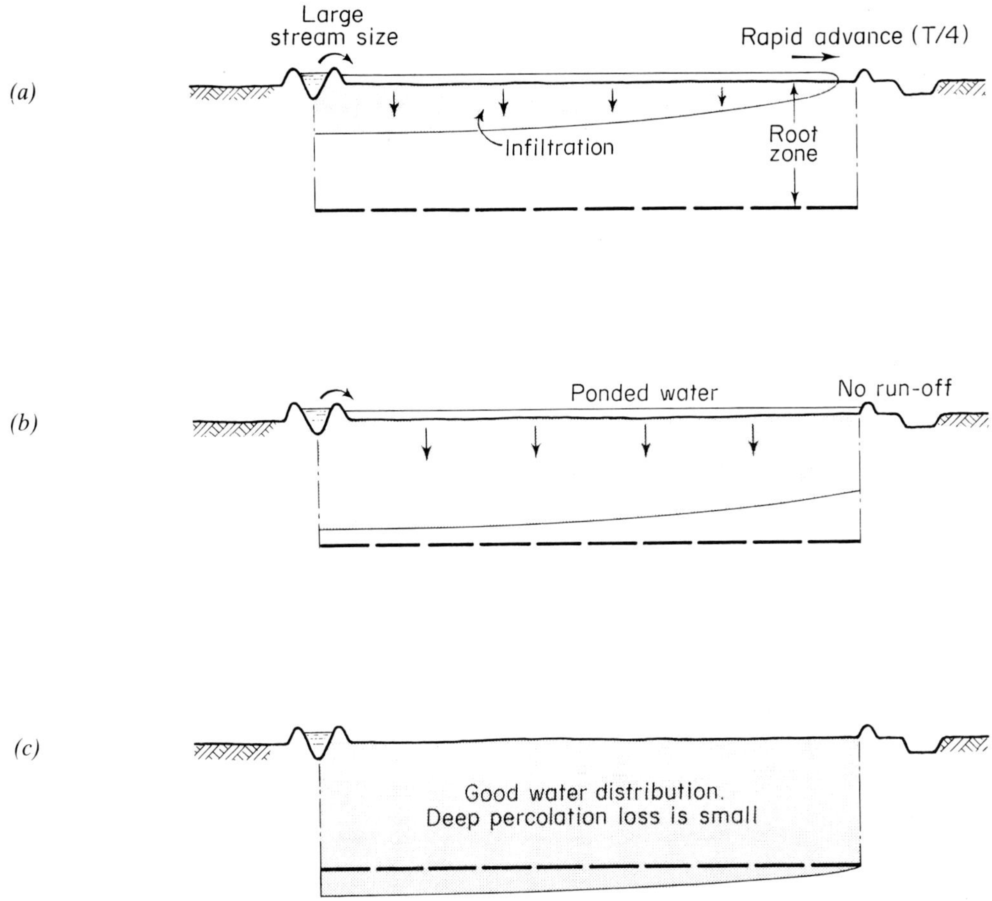

2.13 Basin irrigation. (a) Rapid advance – very little infiltration takes place (b) Water ponded on surface – infiltration occurs over entire basin (c) Completed irrigation

Basin irrigation

A uniform distribution of water is obtained in basins by using a stream size large enough to spread water quickly across the soil surface (figure 2.13). This reduces the difference in contact time between the top of the field and the far end and so deep percolation losses are kept small. Just how large the stream size must be depends on the irrigation depth, soil type and basin size. A practical guideline is to use a stream size large enough for the flow to reach the far end of the field in one quarter of the contact time. This is called the *quarter time rule*. For example, if the contact time required to fill the soil reservoir is 50 minutes, water must reach the far end of the field in 50/4 = 12.5 minutes after starting the irrigation. Using this method deep percolation losses will be less than 10% of the total water applied.

When irrigating basins, run-off is not usually a problem. The land surface is level and so water can be ponded and left to infiltrate into the soil. Basin irrigation practice is described in more detail in chapter 3.

Border irrigation
A uniform distribution of water is obtained in border irrigation by using a small stream size and stopping the inflow at the right time (figure 2.14).

The small stream size means that water advances slowly along the field. When sufficient water has infiltrated at the top of the field the inflow is stopped. The flow recedes away from the farm channel stopping any further infiltration. This part of the irrigation is called the *recession*. Further along the field there is still a lot of water on the soil surface and so the advance and infiltration continue. All the surface water is gradually absorbed at the far end of the field.

This method can result in a very uniform irrigation but in practice it is very difficult to achieve. It depends on choosing the right stream size for the soil type, irrigation depth and length of border and stopping the inflow at the right time. If the inflow is stopped too soon water may not even reach the end of the border. If it is stopped too late then deep percolation and run off can occur.

Some irrigators use practical guidelines such as stopping the inflow when the advance reaches two thirds of the length of field. However, rules like this do not always work well because water advances at different rates on different crops, soils and land slopes. It can also differ from one irrigation to another in the same field because the roughness of the field changes as the crops grow. Once the irrigation system has been built the irrigator must decide for himself on the best rules and practices to use on his fields.

Border irrigation practice is described in more detail in chapter 4.

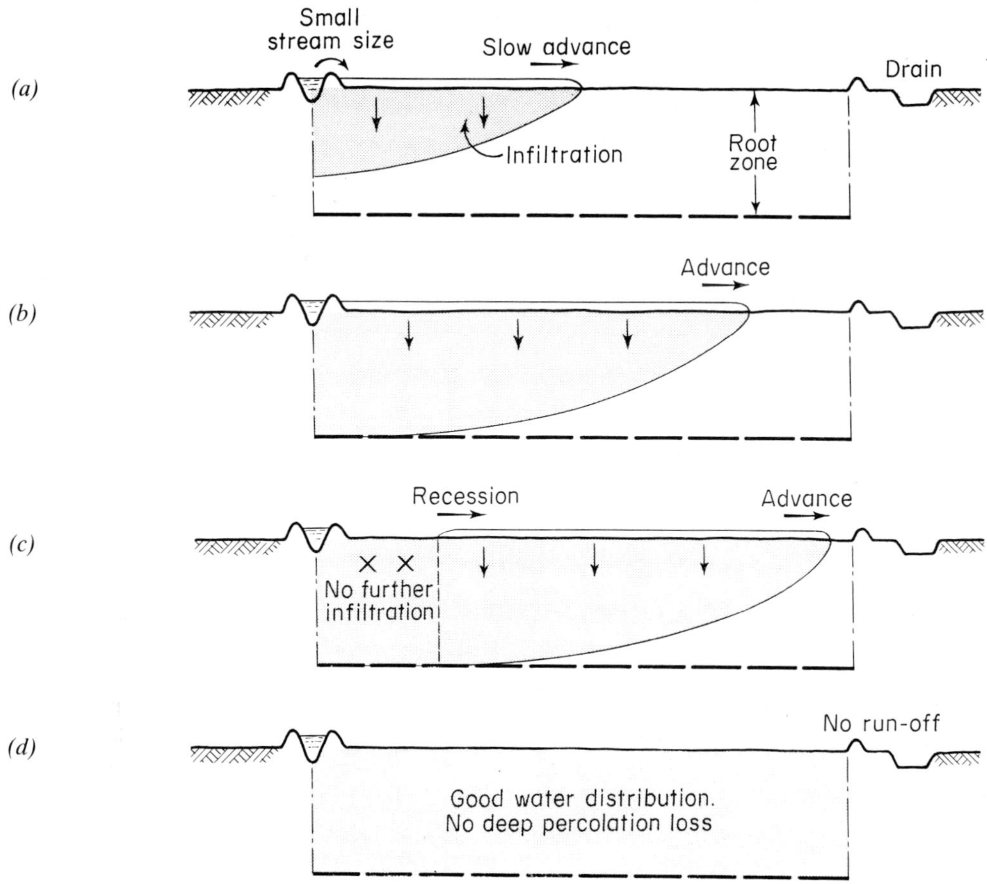

2.14 Border irrigation. (a) Advance begins (b) Inflow stopped (c) Advance continues and recession begins (d) Irrigation completed

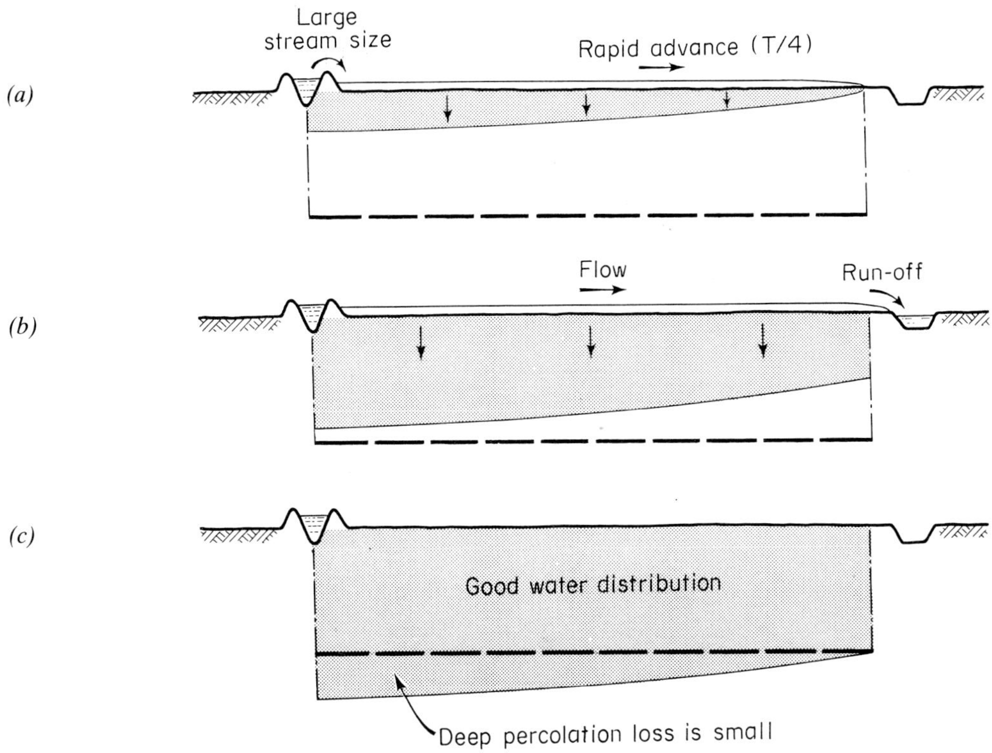

2.15 *Furrow irrigation. (a) Rapid advance – very little infiltration takes place (b) Flow continues – water runs off end of furrow (c) Irrigation completed*

Furrow irrigation

Furrow irrigation in some ways is very similar to basin irrigation. A large stream size is used to spread water quickly along each furrow to keep deep percolation losses small (figure 2.15). The size of stream required depends on the irrigation depth, soil type and furrow length. The quarter time rule is often used as a practical guideline for the advance of water down furrows. Unlike basins, however, run off can be a problem. When the flow reaches the far end of the field it cannot be easily ponded because furrows are usually on sloping land and there is very little room to store water. Water that cannot infiltrate fast enough into the soil must be allowed to run off the end of the field and into a drain. This may seem to be a great waste of water but there are ways in which this loss can be reduced. This is discussed in more detail in chapter 5.

An experiment

It is obviously not possible to see water infiltrating into the soil during field irrigation but the reader may like to make a simple demonstration apparatus. The following equipment is needed:

- Two sheets of 4 mm glass 600 mm × 400 mm
- Three pieces of wood 30 mm × 20 mm, 380 mm, 600 mm and 200 mm long
- Two laboratory stands and clamps (rubber covered)
- One bucketful of fine sand
- 2 m of 9 mm diameter plastic tubing
- One funnel – for inflow
- Water supply (tap or beaker full of water)
- Measuring flask – for collecting run-off

The sheets of glass are clamped and held apart by the pieces of wood to make a small tank (figure 2.16). This is filled with sand, the surface is smoothed and a line drawn on the glass to represent the root zone. Water is then poured onto the sand at one end to represent

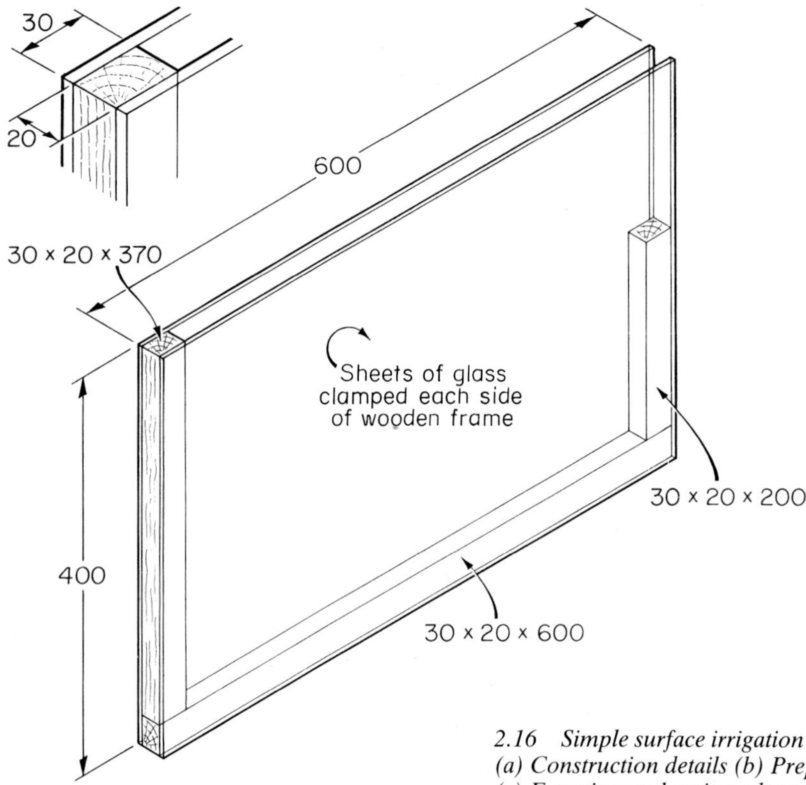

(a)

2.16 *Simple surface irrigation demonstration.*
(a) Construction details (b) Preparing experiment
(c) Experiment showing advance and infiltration

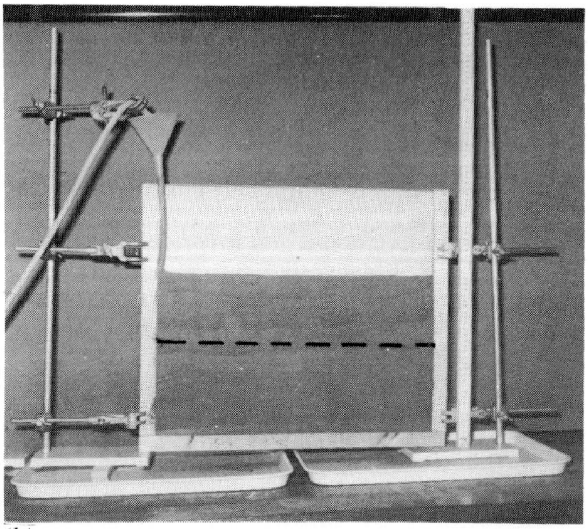

(b)

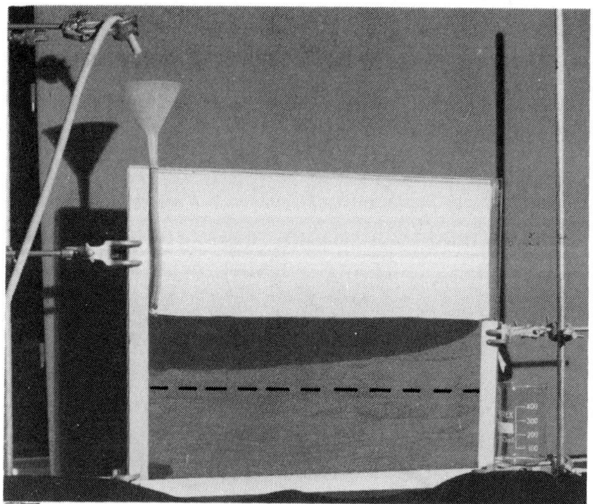

(c)

the inflow from a farm channel. The water should move along the sand surface and infiltrate in the same way as it does in surface irrigation. The reader can then experiment for himself and discover the many factors which affect the distribution of water. After each experiment the sand must be emptied and dried.

Try the following:

Use different stream sizes. For a small stream size the advance may not even reach the far side of the tank and a lot of water is lost from deep percolation (figure 2.16). For a large stream size the water may reach the far end of the field and start to run off before it has had time to infiltrate properly

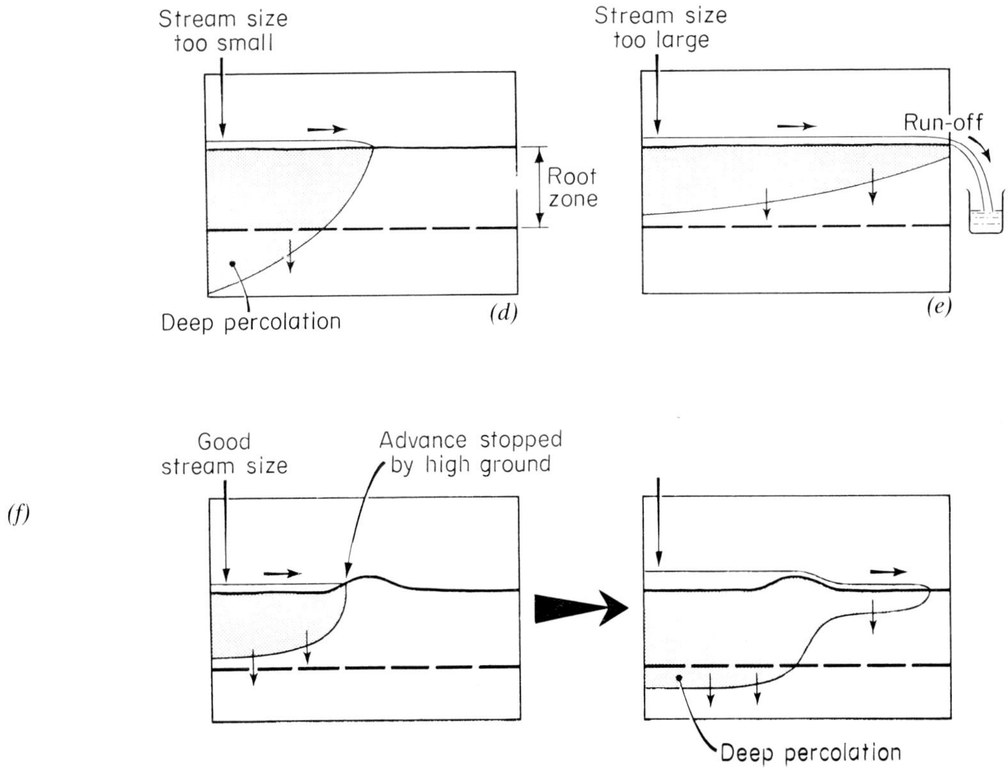

2.16 (d) Effects of small stream size (e) Effect of large stream size (f) Effect of high ground in a field

(figure 2.16). Run off must continue if all the root zone is to be properly filled.

Cut small hollows or humps into the soil surface to see how important land preparation is for uniform water distribution (figure 2.16).

Use different soil types, for example a sand and a sandy loam soil. (Soils with clay in them will be too messy to handle). Using the same stream size for each soil observe how the advance and infiltration change.

There are many other interesting experiments that can be done with this simple tank. The reader might like to experiment further on his own.

Conclusion It would seem that surface irrigation can be quite complicated if an irrigator takes notice of all the factors which affect the movement of water on the surface and in the soil. This may be surprising since surface irrigation is usually thought of as the simplest of irrigation methods. It is in fact very easy to practice surface irrigation but it is very difficult to do it well. To irrigate uniformly and efficiently requires considerable skill.

2.12 Useful formula

To irrigate properly it is important to know the following:

Irrigated area

Irrigation depth

Stream size

Irrigation time

Irrigated area
This is the area of the field or farm measured in hectares (ha) which is to be irrigated (figure 2.17).

Irrigation depth
The amount of irrigation water is generally referred to by depth rather than volume. This is the depth of water that must infiltrate into

the soil to fill the soil reservoir. It is measured in millimetres (mm). Measuring the irrigation depth is beyond the scope of this book. However as an approximate guide a crop can use up to 5 mm/day in a humid region and 8-12 mm/day in an arid region.

Irrigation time and stream size
The irrigation time is the time in hours (h) required to supply the irrigation depth (mm). It depends on the stream size measured in litres per second (ℓ/s). If the stream size is known the irrigator can calculate the irrigation time. If the time is limited then the stream size required to complete the irrigation can be calculated.

If any three of the above factors are known the fourth can be calculated using the formula:

$$\text{Stream size } (\ell/\text{s}) = \frac{2.78 \times \text{irrigation depth (mm)} \times \text{irrigated area (ha)}}{\text{irrigation time (h)}}$$

Example
An average depth of 100 mm of water is needed to irrigate 5 ha of land in 20 hours. What stream size is needed?

Using the above formula:

$$\text{Stream size } (\ell/\text{s}) = \frac{2.78 \times 100 \times 5}{20}$$

$$= \underline{69.5\ \ell/\text{s}}$$

If the stream size available is only 40 ℓ/s what would be the new irrigation time?

Re-arranging the above formula:

$$\text{Irrigation time (h)} = \frac{2.78 \times \text{irrigation depth (mm)} \times \text{irrigated area (ha)}}{\text{stream size } (\ell/\text{s})}$$

$$= \frac{2.78 \times 100 \times 5}{40}$$

$$= \underline{34.75\ \text{h}}$$

Always make sure to use the right units of measurement in the formula.

2.13 Efficiency
This is a term used to describe all the water losses in surface irrigation. It describes the water required by the crop in relation to the amount of water applied. It is usually expressed as a percentage (%).

$$\text{Irrigation efficiency } (\%) = \frac{\text{water required (mm)}}{\text{water supplied (mm)}} \times 100$$

It is unlikely that an irrigator would achieve 100% efficiency. This would mean a completely uniform irrigation with no wastage. Most obtain between 50-90% but some are much lower than this. A great deal depends on the irrigation system and the skill of the irrigator.

Low efficiencies can mean reduced crop yields because of water logging. It may also mean that if some fields are receiving too much water then other fields may not be getting enough.

Example
75 mm of water is needed to irrigate a field. In order to irrigate adequately over the whole area an average depth of 100 mm must be applied. How much water is lost? What is the irrigation efficiency.

$$\text{Water loss} = \text{water depth supplied (mm)} - \text{water depth required (mm)}$$

$$= 100 - 75$$

$$= \underline{25\ \text{mm}}$$

$$\text{Irrigation efficiency} = \frac{\text{water required (mm)}}{\text{water supplied (mm)}} \times 100$$

$$= \frac{75}{100} \times 100$$

$$= \underline{75\%}$$

2.14 Clay soils
In many countries very heavy clay soils are irrigated with surface methods. They have

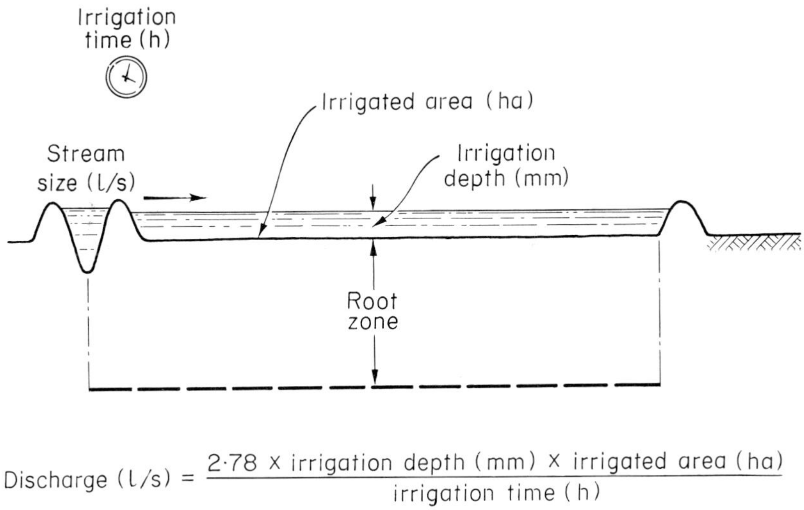

2.17 Useful formula.

very low infiltration rates and large, deep cracks appear when the soil dries.

These soils can be irrigated easily but water does not move over the land surface and infiltrate as described in 2.10. Instead it runs down the cracks in the soil. Gradually the soil swells as water is absorbed and the cracks close stopping any more water from entering. Very high irrigation efficiencies can be achieved because there is no deep percolation loss. However, the soils are very weak when they are wet and machines cannot drive on them. They are good for irrigation but difficult for mechanisation.

2.15 Salinity

Although deep percolation is normally thought of as wasting water, in arid regions it can be of benefit.

Rivers and streams usually contain small quantities of salt dissolved in the water. When this is used for irrigation, the water is taken by the plants but the salt is left behind. Salt can damage and even kill plants if there is too much. In extreme cases the whole ground surface becomes covered in white salt. On some schemes there is so much salt that people collect it and use it for cooking.

To overcome this problem it is common practice to apply more water than is needed by the crop. The excess water dissolves the salt as it moves down through the soil and carries it beyond the root zone. Pipes are usually buried in the soil below the root zone to remove this water and dispose of it safely in drains. Washing salts from the soil in this way is called *leaching*.

3

BASIN IRRIGATION

3.1 Introduction

Basins are the simplest and most widely used of all surface irrigation methods. A field may be divided into one or more basins. Each one is a level area of land surrounded by earth bunds in which water can be ponded until it infiltrates into the soil (figure 3.1).

Basin irrigation can be adapted to suit many crops, soils and farming practices. However, it is important to use the right size and shape of basins and good water management for the method to work well.

3.2 Size

There are many different sizes of basins. Some are only 1-2 m^2, while others are 3-4 ha or more. The size depends on:

- Soil type
- Stream size
- Irrigation depth
- Field size
- Land slope
- Farming practice

Soil type, stream size and irrigation depth

These three factors are most important in determining the area that can be enclosed in each basin (figure 3.2). To understand this it is useful to remember the way in which basins are normally irrigated, (section 2.11).

When irrigating sandy soils water infiltrates rapidly. This means that basins must be small for water to spread quickly even when large stream sizes are used.

When irrigating clay soils water infiltrates much more slowly and so there is more time for water to spread over the soil surface. This means that basins can be much larger than on

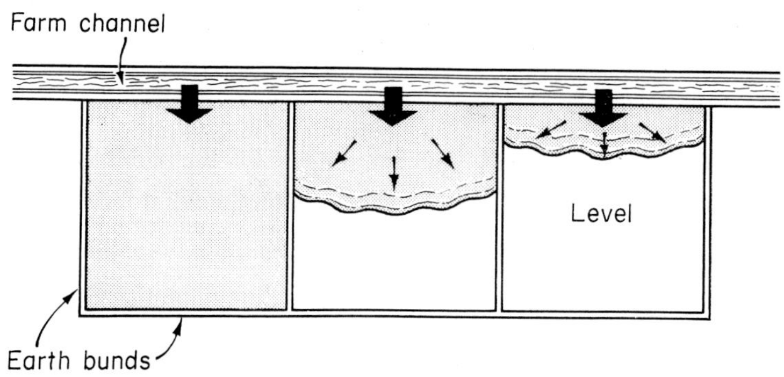

3.1 Basin irrigation.

sandy soils even when the stream size is small.

Basins can usually be much larger on the same soil when a larger stream size is available. This is because water will spread more rapidly across the soil surface.

Applying a larger irrigation depth also means that basins can be increased in size. A larger depth of water needs a longer contact time and so there is more time available for water to spread across the soil surface.

3.2 Factors affecting basin size.

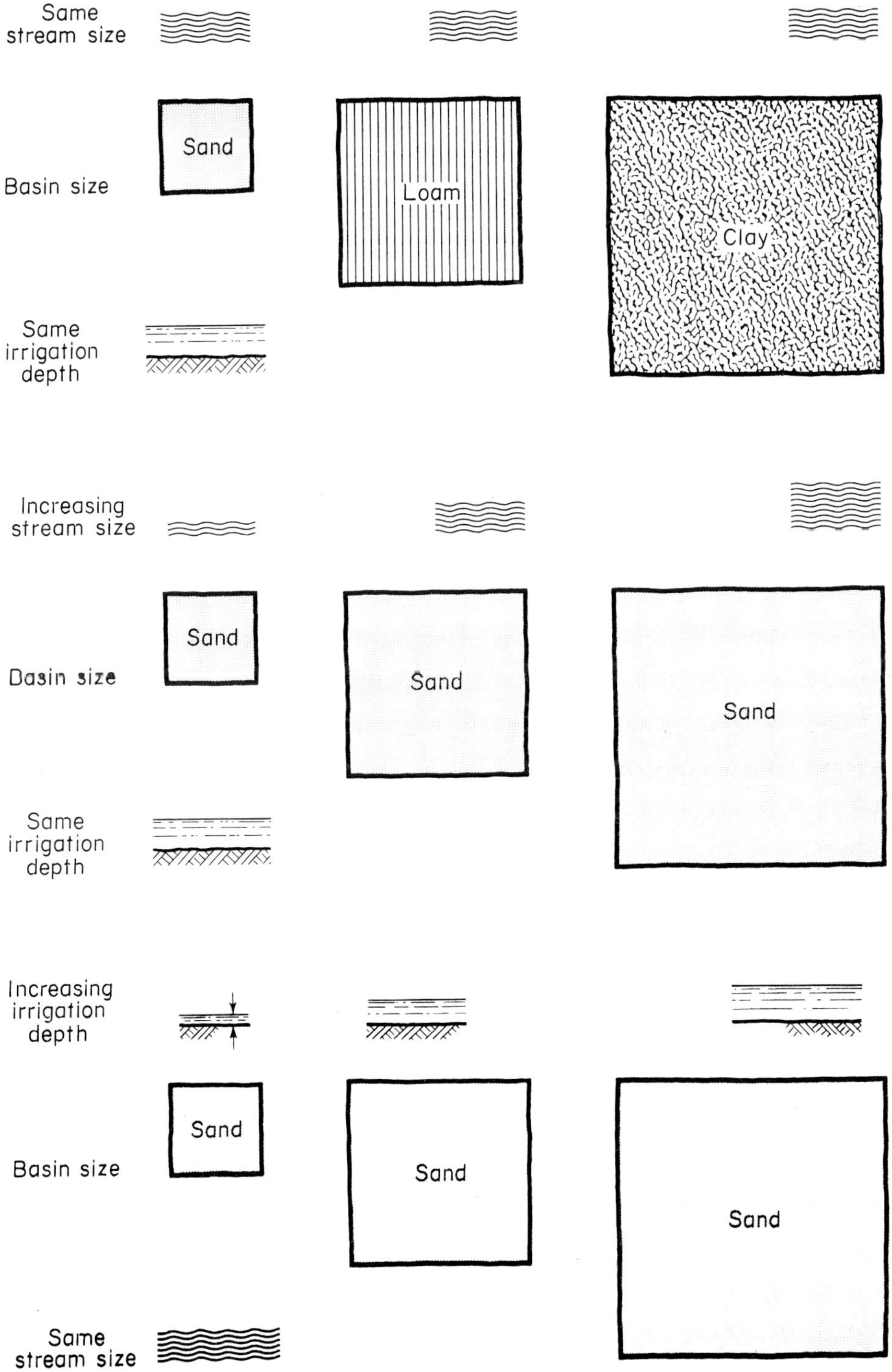

(a) Basins can be larger on clay soils than on sands

(b) Basin size can be increased as stream size is increased

(c) Basin size can be larger when a larger irrigation depth is applied

There are no simple calculations to help an irrigator to select the best basin sizes for different stream sizes, irrigation depths and soil types. In places where basin irrigation is practiced, the experience of local irrigators often provides a good guide to the best size. When a new scheme is being planned or the local irrigation practices are obviously poor, the experience of others irrigating in similar conditions is needed. Table 3.1 is a collection of such experience. It provides a good guide to basin sizes for different stream sizes and soils which can give a uniform and efficient irrigation.

For example, (from table 3.1) when the stream size is 30 ℓ/s the basin size is only 0.02 ha on sandy soil. The basin can be increased to 0.2 ha using the same flow on a clay soil. Notice also how large the stream sizes must be even for the smallest basins.

To summarise: (figure 3.2)

Basins can usually be much larger on clay soils than on sandy soils.

Basins can be larger when a larger stream size is used for irrigation.

Basins can be larger when a larger irrigation depth is applied.

Table 3.1 Suggested basin sizes (ha)

Stream size (ℓ/s)	*Soil type* *Sand*	*Sandy loam*	*Clay loam*	*Clay*
15	0.01	0.03	0.06	0.1
30	0.02	0.06	0.12	0.2
60	0.04	0.12	0.24	0.4
90	0.06	0.18	0.36	0.6
120	0.08	0.24	0.48	0.8
150	0.10	0.30	0.60	1.0
180	0.12	0.36	0.72	1.2
210	0.14	0.42	0.84	1.4
240	0.16	0.48	0.96	1.6

Field size

This may limit the choice of basin size. In very small fields the basin may be the same size of the field. In large fields a common practice is to divide the land into basins of equal size and shape. This makes it easier to deliver the same amount of irrigation water to each basin.

3.3 Effect of land slope on basin size and shape. (a) Large rectangular basins on flat land (USA)

Land slope

As the soil surface within each basin must be level, basin size can depend on the land slope (figure 3.3).

When the land is level, basins can be as large as the stream size and soil type will allow.

On sloping and undulating ground the land surface must be re-shaped into level areas. Figure 3.3[b] is an example of basins used on very steep land to grow rice. A 'staircase' has been constructed to provide the level land for the basins. The steps are called *terraces* and normally the drop in level between them should be less than 150 mm to avoid erosion problems (section 3.8). In such cases the basin size is determined by the size of the terrace. This in turn is limited by the land slope and the depth of fertile top soil (figure 3.4). When the top soil is very deep the maximum drop of 150 mm limits terrace width. When the top soil is shallow there is the danger of exposing unfertile sub-soils if there is too much excavation. The terrace width is limited in this case by the depth that can be excavated. Suggested terrace widths for different land slopes are given in table 3.2. The smaller width is for areas with shallow top soil, the larger width is for areas with very deep top soil.

Farming practice

In many developing countries farms and fields are small, often only 1-2 ha or less. A wide range of crops are often grown at the same

3.3(b) Contour basins for rice on sloping land (Sri Lanka)

Table 3.2 Suggested terrace widths for basins

Land slope (%)	*Terrace width* (m)
0.1	150-60
0.2	75-30
0.5	30-12
1.0	15-6
1.5	10-4
2.0	7.5-3
3.0	5-2
4.0	3.75-1.5

time and cultivation, planting and harvesting are done by hand or by animal power. Small basins are often used for this type of farming. They are easily levelled by hand; they divide up the small fields for planting and they require only small stream sizes to irrigate them.

On large mechanised farms, channels and earth bunds surrounding basins are obstacles for the machines. Basins need to be as large as possible so that machines can turn round easily and have long runs without too many turns (figure 3.5). The basin dimensions also need to be some multiple of the machine width. For example, if a cutter blade 2 m wide is used to harvest alfalfa the basin width should be some multiple of 2 m.

Sometimes the type of cropping limits basin size. For example a small basin can be used to irrigate single trees in an orchard (figure 3.6) or a small plot in a vegetable garden.

3.3 Shape

The shape of basins is mainly determined by the land slope.

When the land is level or sloping uniformly

basins can be rectangular (figure 3.7). This makes it easier to plan the layout of canals, drains and roads around the farm and to use machines in the field.

When the land is undulating, basins can be shaped to the land contours. These are called *contour basins* and they can be very irregular in shape (figure 3.3[b]). On some schemes undulating land is re-graded into large flat areas so that rectangular basins can be used (see chapter 7).

Rectangular basins are generally long and narrow with the shorter side located alongside the farm channel (figure 3.7). This reduces the number of farm channels required which in turn reduces labour and maintenance costs and improves vehicle access to the fields. On some farms it is possible to irrigate basins on both sides of the farm channel which further reduces the number of channels.

3.4 Crops

Many different crops are grown in basins. These include field crops such as rice, alfalfa and cereals; row crops such as cotton, maize and groundnuts, and orchard crops. Rice is perhaps the most common basin crop. The level land on which water can be ponded provides the ideal growing conditions for this crop.

3.3(c) Rectangular basins for vegetables on sloping land (Vietnam)

In orchards, basins can be easily adapted to the needs of growing trees. When young trees are first planted the root system is small and water demand is low. A small basin around each tree is sufficient (figure 3.6). As the trees grow and their root systems expand the basins can be increased in size to supply more water.

Some crops suffer when they have to stand for long periods in water or very wet soil. These should not be grown in level basins but on furrows (see chapter 5). An alternative is to grow them on raised beds or furrows within basins. Vegetables are often grown in this way (figure 3.8). This method is also used when the soils are very heavy and water may stand in the basins for 24 hours or more before infiltrating.

3.5 Earth bunds

These are small earth embankments or dams built around each basin to hold water without leakage while it infiltrates (figure 3.9). The size and shape of bunds depends on:

- Irrigation depth
- Freeboard
- Wave action
- Farm machinery

Bunds are usually built 150-300 mm high to contain water on the soil surface (irrigation depth 50-200 mm) with a small freeboard to stop water flowing over the top. In large basins waves can be a problem in windy conditions. In such cases a larger freeboard will be required (figure 3.9).

Bunds vary between 0.6-1.2 m wide at the base. This provides sufficient soil to stop leakage.

Bunds in rice fields are usually much larger than for other crops. They are 400-500 mm high and 1.5-1.8 m wide at the base. This is because rice is often grown as a single crop on the same land and so more permanent bunds are constructed. On some farms the bunds are used as paths to the rice fields.

On mechanised farms bunds are carefully formed so that farm machinery can be driven over them easily (figure 3.9).

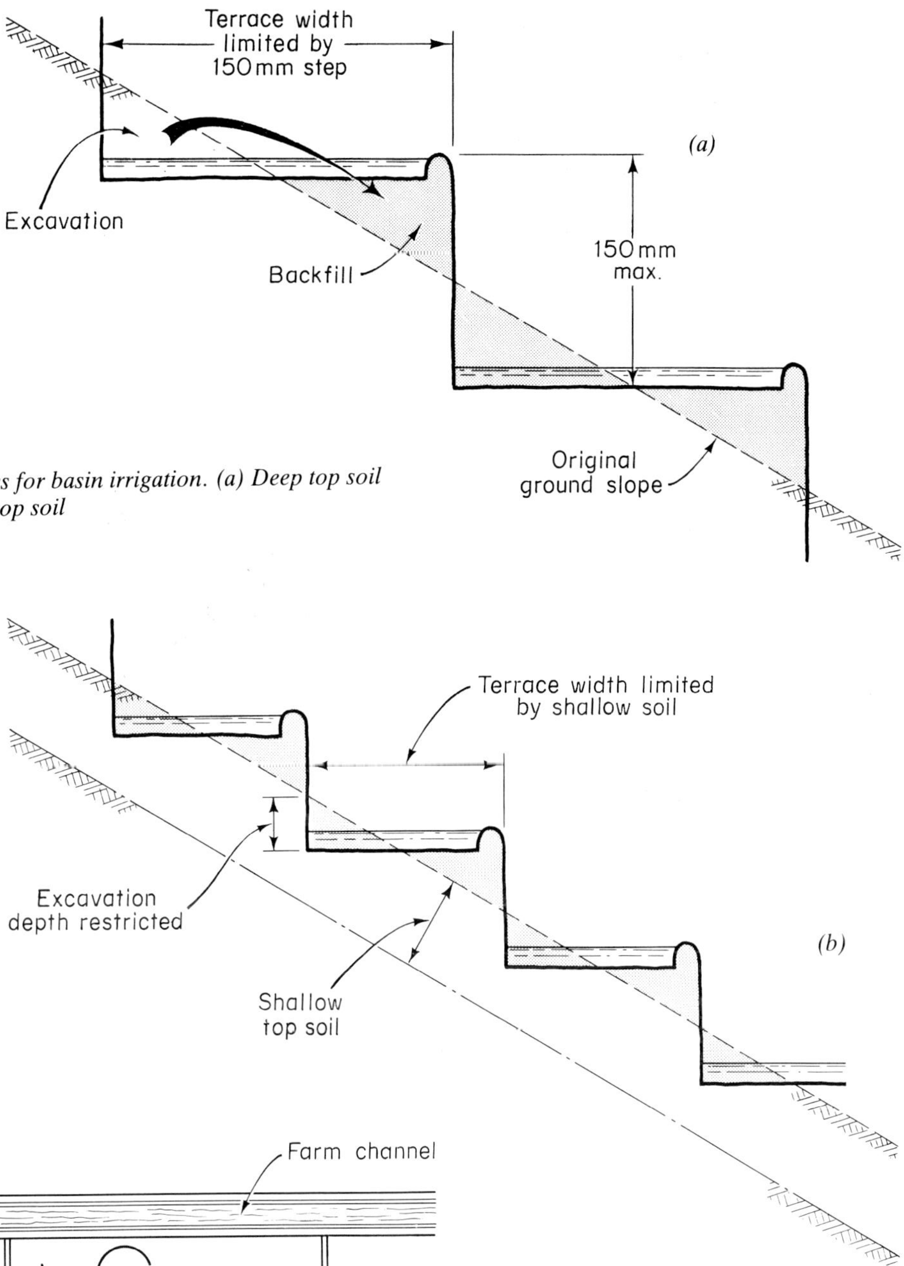

3.4 Terraces for basin irrigation. (a) Deep top soil (b) Shallow top soil

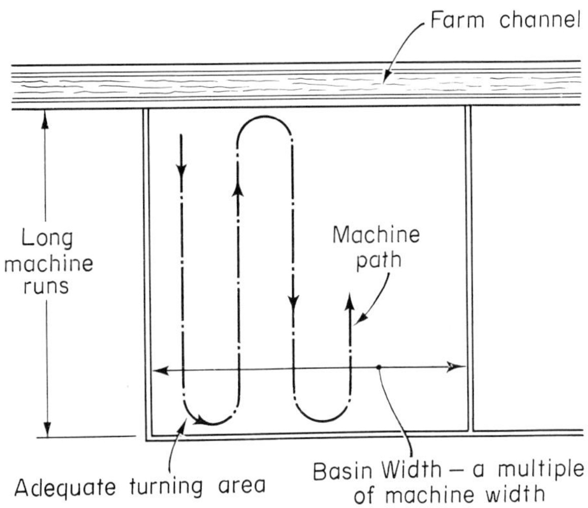

3.5 Large basins used for mechanised farming.

3.6 *Small basin for a single tree.*

Construction

The following steps are taken when constructing bunds:

Setting out

Collecting soil

Shaping

Land smoothing

Before construction can begin the position of each bund is set out on the ground. This is done with small flag markers or by spreading white powder such as chalk along the line.

Collecting the soil to form the bund can be done by hand or by a tractor drawn disc plough. The soil is usually taken from an area close to the bund and this forms a borrow-furrow. Crops do not grow well in the borrow furrow and it tends to remain very wet (figure 3.9).

3.7 *Rectangular basins. (a) General view of basins on a large scheme (Iraq)*

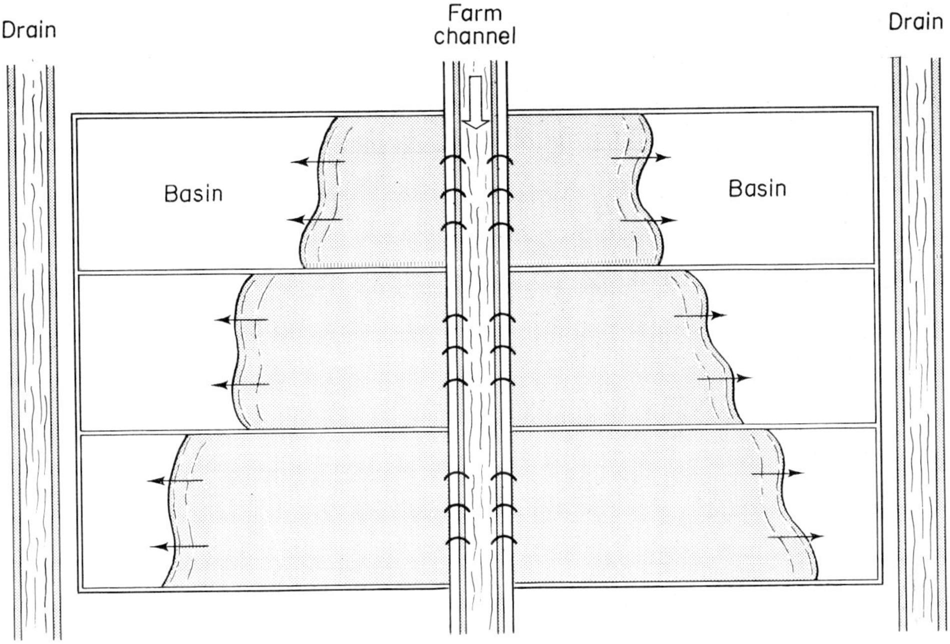

3.7(b) Typical layout

3.8 Furrow-basin irrigation for vegetables.

3.9 Earth bunds. (a) Large bunds used for access in rice fields

A special ridging machine is available to overcome this problem. The machine collects soil from a wider area producing a much wider and shallower borrow-furrow in which crops can grow better.

Once the soil has been collected it can be shaped with a tractor mounted A-frame (figure 3.10) and compacted with a roller. It is usually built higher than is normally required to allow for settlement when the loose soil is wetted.

The final step is to smooth out any differences in level in the basins caused by taking soil to form the bunds. This can be done by hand in small basins or by a land plane in larger basins (see chapter 7).

For most annual crops bunds are formed temporarily each season. For more permanent crops larger permanent bunds are formed but these need regular maintenance each season.

WATER MANAGEMENT

3.6 Irrigating basins

When irrigating basins, water from the farm channel is turned into each basin using spiles or siphons. On some farms all the flow is directed into each basin in turn. On others the flow may be divided to irrigate several basins at the same time.

Whatever method is used the stream size must be large enough for the flow to advance rapidly across the basins. This will ensure a reasonably uniform irrigation, see page 19. The basins are then filled to some desired irrigation depth and the water ponded until it infiltrates into the soil.

Some water will always be lost in deep percolation when irrigating basins but this will be quite small if the right stream size is used. There should be no run-off losses because water can be ponded on the level soil surface.

3.7 Methods of supplying water

There are two ways of supplying water to basins:

Direct supply

Cascade supply

Direct supply

This is the best method for most crops and soils. A farm channel is constructed alongside every basin in the field (figure 3.7). Each basin is supplied directly from the farm channel.

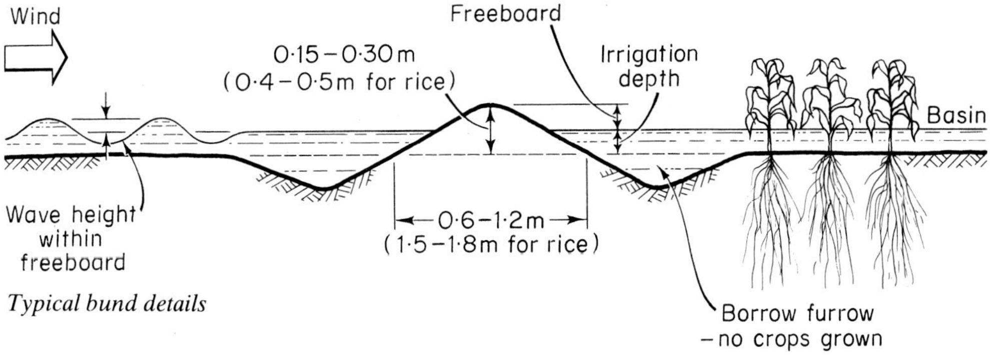

3.9(b) Typical bund details

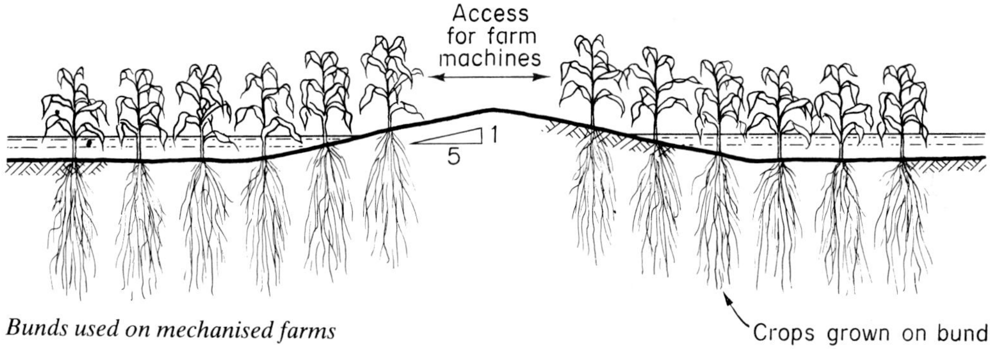

3.9(c) Bunds used on mechanised farms

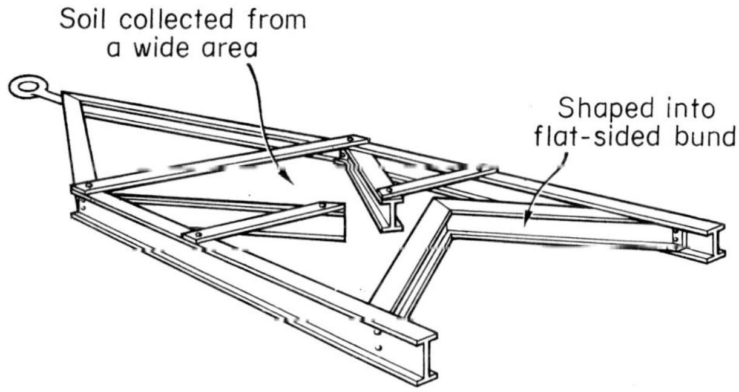

3.10 Tractor mounted 'A' frame for forming bunds.

Cascade supply

This method is sometimes used on sloping land where basins are constructed on terraces (figure 3.11). Water is supplied to the first basin directly from the farm channel. It then flows from one basin to the next down the slope. There are two ways of using a cascade.

The first way is to fill basin 1, then basin 2 and so on until all the basins are irrigated. This is just the same as irrigating one large basin but in this case it is divided up into smaller ones, at different levels. To irrigate efficiently water must reach the last basin in one quarter of the contact time. The method is only suited to clay soils with low infiltration rates. On sandy soils deep percolation losses in the top basins can be very high. This method is often used for rice irrigation. Small structures are built into the bunds to control the flow from one basin to the next. A common practice when irrigating rice is to release a small continuous stream into the cascade throughout the season. This is adjusted regularly as the water demand of the crop changes. The irrigator checks that sufficient water is entering the cacade by ensuring that there is always a small excess flowing into the drain.

The second method is more suited to soils with higher infiltration rates. The stream is released into a small channel alongside the upper basins 1, 2 and 3 so that basin 4 is irrigated first (figure 3.11). The bunds between basins 4 and 3 is then closed and basin 3 is irrigated. Basin 1 is the last one to be filled. This is very similar to the direct method but use is made of the small borrow furrow to carry the flow instead of constructing more farm channels.

3.8 Soil erosion

When irrigating with large stream sizes on sloping land there is always the risk of soil erosion. If too much water is released into the basins at the top of a cascade system it may flow out of control over the earth bunds causing erosion further down. A serious flood could wash away the top soil and damage fields beyond repair. Great skill is needed to irrigate in these conditions.

(a)

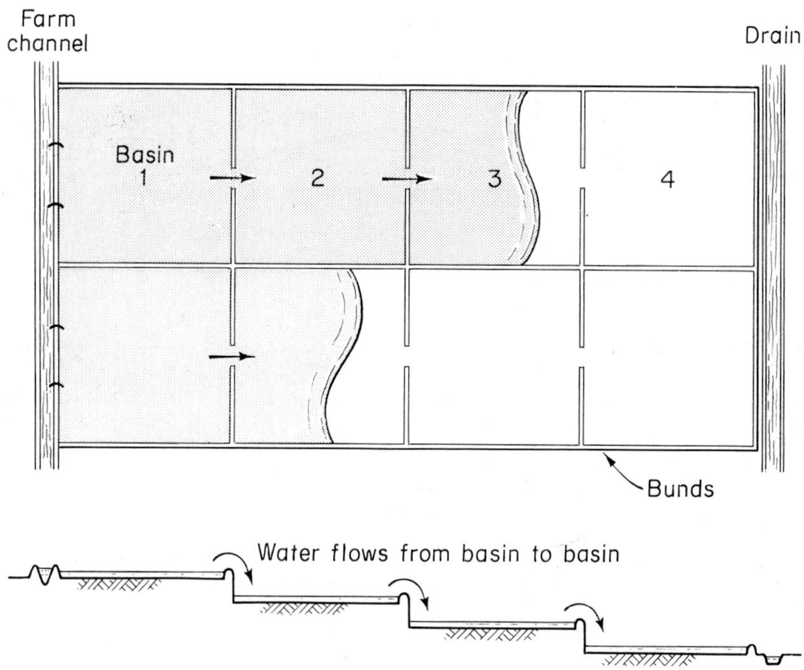

(b)

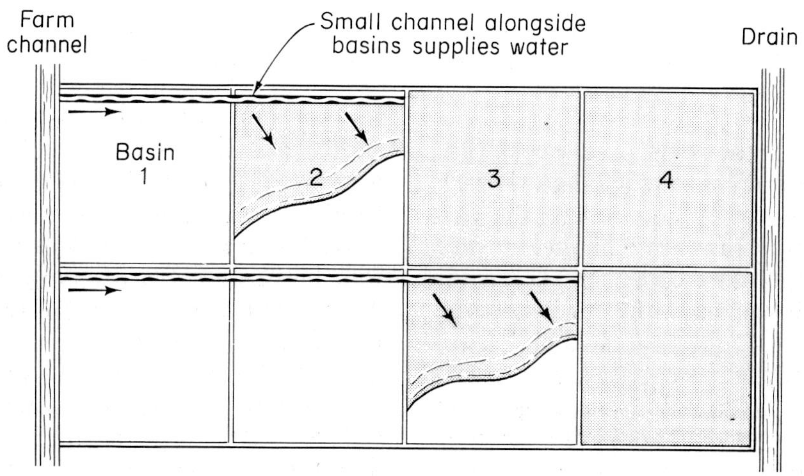

3.11 'Cascade' method of irrigating basins. (a) For clay soils (b) For sandy soils

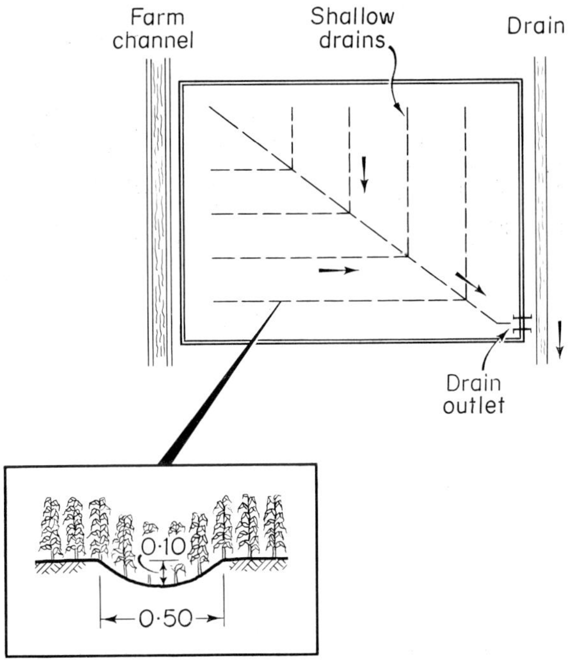

3.12 *Shallow drains in rice basins.*

3.9 Leaching

Basin irrigation can be a very useful and simple method of leaching harmful salts from the soil because water can be ponded for long periods (section 2.15). The basins are filled with water to some depth greater than the normal irrigation depth and left until it infiltrates into the soil. The excess water percolates down through the root zone washing out the harmful salts.

3.10 Drainage

Drains are needed to remove excess water from basins. This may be due to excess irrigation water supplied from the farm channel or the result of a heavy rainstorm. When a rainstorm is exceptionally heavy, water is sometimes held in the basins and slowly released into the drains to avoid overloading the channels and flooding downstream. However, it is important that water does not stand in the basins for longer than a few days as some crops may be damaged and yields reduced. This of course does not apply to rice which grows best in flooded conditions.

Shallow drain channels are sometimes dug in large basins to help in draining them. This is often done in rice basins so that they can be drained quickly, at the end of the season for harvesting. A 'herring bone' pattern of small channels guide the flow towards the drain (figure 3.12).

3.11 Common faults

Although basin irrigation can be very efficient there are common irrigation practices which result in poor uniformity and low irrigation efficiency. Some of these are described below.

Poor land preparation

When a basin is constructed on sloping ground and the land is not levelled, too much water ponds at the lower end (figure 3.13) and is wasted in deep percolation (see example). Crops planted at the lower end may also suffer from water logging and reduced yields.

If a basin is constructed on reasonably level ground but the land has not been levelled properly too much water lies in the low areas and not enough is applied to the higher ground. Plants in the low areas may suffer from too much water while those on the high ground suffer from too little. Both result in poor crops.

These faults can be corrected by careful land preparation (chapter 7).

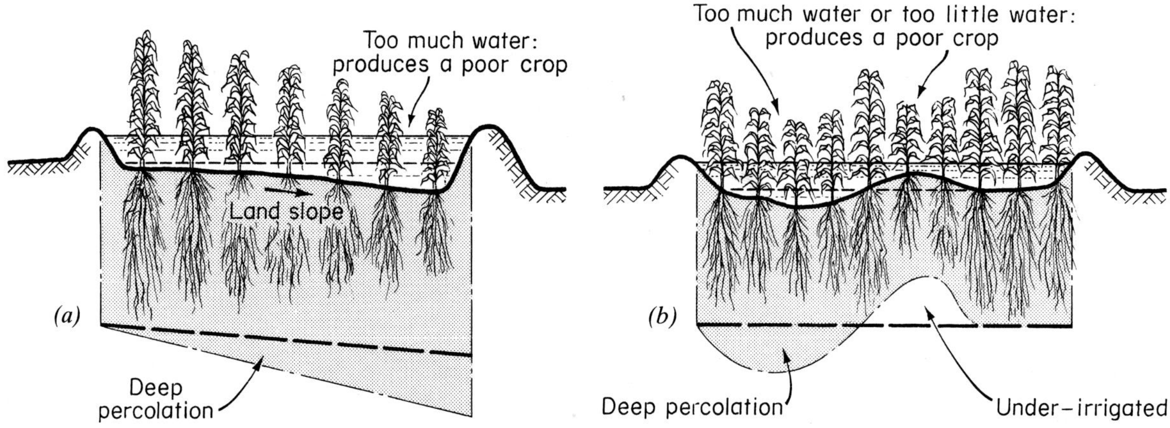

3.13 Common fault – poor land preparation. (a) On sloping land (b) Flat land

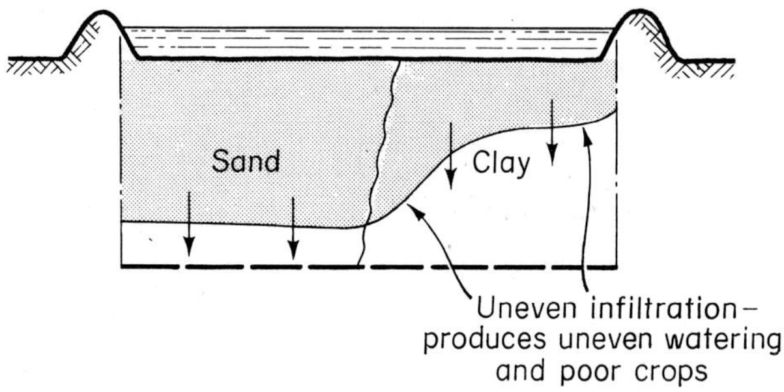

3.14 Common fault - basin covering more than one soil type.

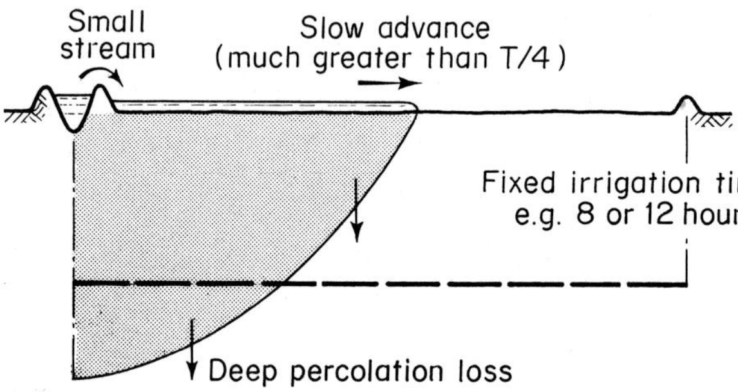

3.15 Common fault – irrigation to a fixed time schedule.

Basins covering more than one soil type
Variations in soil type within a basin can result in water infiltrating much faster in one place than in another. On the higher infiltration rate soil water may be lost from deep percolation. On the lower infiltration rate soil insufficient water may be applied (figure 3.14). The only solution is to realign basin boundaries so that each basin contains only one soil type.

Fixed time schedule
Irrigation time is sometimes fixed (eg 8 or 12 hours) so that changing the flow from one basin to the next fits in with other work on the farm. This often means using a small stream size. The quarter time rule is ignored, the advance is too slow and a lot of water can be lost through deep percolation (figure 3.15). Many irrigators are not aware of deep percolation losses, because it is below ground and cannot be seen.

Example
A basin 10 m wide is constructed on sloping ground and the land is not levelled (figure 3.16). The irrigation depth required to fill the soil reservoir is 100 mm. What irrigation depth must actually be applied? How much water will be lost? What is the irrigation efficiency?

To apply 100 mm of water at the upper end, 150 mm must be applied at the lower end.

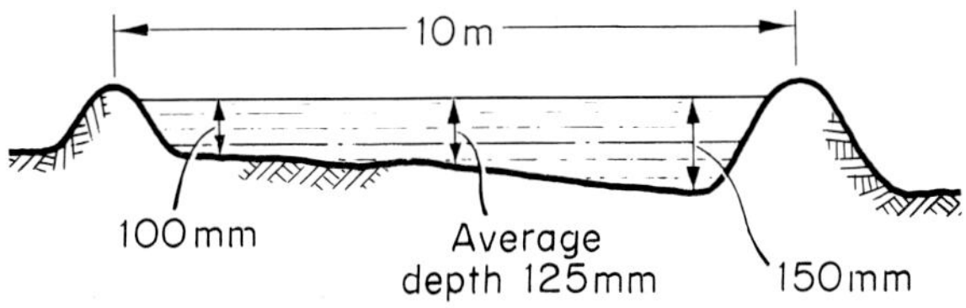

3.16 Basin on sloping ground (see worked example).

$$\text{Average irrigation depth} = \frac{100 + 150}{2}$$

$$= 125 \text{ mm}$$

This is the depth of water that must be supplied to ensure the basin is irrigated adequately.

Water loss = water depth supplied − water depth required

$$= 125 - 100$$

$$= 25 \text{ mm}$$

$$\text{Irrigation efficiency} = \frac{\text{water depth required}}{\text{water depth supplied}} \times 100$$

$$= \frac{100 \times 100}{125}$$

$$= 80\%$$

Thus a small difference in level of 50 mm across a 10 m wide basin reduces the irrigation efficiency to 80%.

3.12 Efficiency

On well managed basins irrigation efficiency can be as high at 90%. When they are not used properly the efficiency will be much lower. The following are guidelines on the effects of poor practice on irrigation efficiency. The figures should be subtracted from 90%.

Common faults	*Subtract*
Poor land preparation	10-20%
Different soil types in basin	5-10%
Fixed irrigation schedule	10-20%

It is easy to see from this how basin irrigation can be a very inefficient method when it is not managed properly.

4

BORDER IRRIGATION

4.1 Introduction

At first glance, borders seem very similar to basins. The land is divided into strips by small earth bunds to irrigate similar crops and soils (figure 1.1 & 4.1). However, there are two important differences:

- Borders usually slope uniformly away from the farm channel in the direction of water flow. They are not level like basins.
- Borders are irrigated in quite a different way to basins (section 2.11).

Borders are often recognised by their characteristic long, narrow shape, but this can be misleading because basins can also be shaped in this way. They are also called *border-strips, border checks* or *strip checks.* The names are different but they all refer to the same method of irrigation.

Border irrigation can be adapted to suit many field and row crops, soils and farming practices. However, it is important to use the right size and shape of border and good water management for this method to work well.

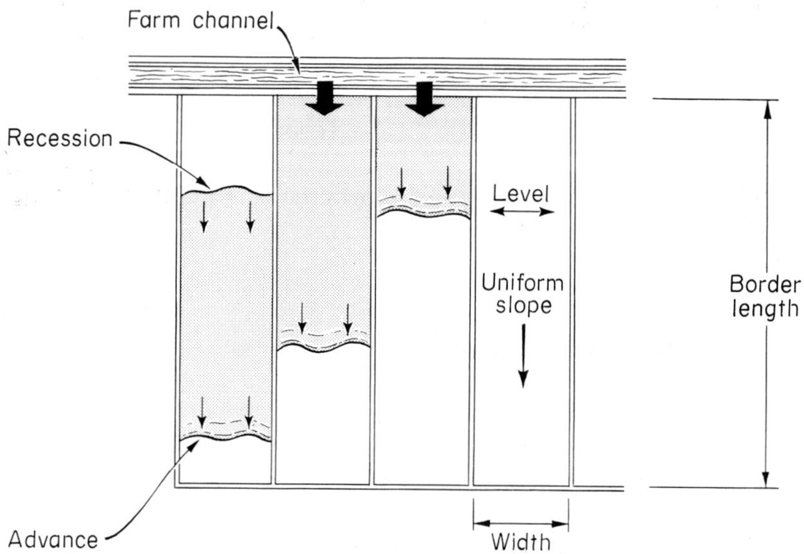

4.1 Border irrigation.

4.2 Size and shape

Borders are usually rectangular in shape, varying in size from 100-800 m long and 3-30 m wide. As a general rule borders should be as long as is practically possible. This means fewer farm channels and drains are needed, less land is taken up by channels and it is much easier to mechanise. Construction and maintenance costs for the channels and structures are also reduced.

There are many factors which affect the choice of border size and shape. These include:

Soil type

Unit stream size

Irrigation depth

Slope

Field size and shape

Farming practice

Soil type, unit stream size, irrigation depth and slope

These factors are the most important in determining the area of land that can be enclosed within each border (figure 4.2). To understand this it is useful to remember the way in which borders are normally irrigated, (section 2.11).

When irrigating sandy soils water infiltrates rapidly. This means that borders must be short (100 m or less) for water to spread down the field even when quite large unit stream sizes are used. If the border was too long for the available stream size, too much water would be absorbed and lost near the farm channel before the flow had time to reach the far end of the border.

When irrigating clay soils water infiltrates much more slowly than in sands and so it can take a long time to fill the soil reservoir. Water cannot be ponded on the soil and left to infiltrate as in basin irrigation and so long borders are needed even when the unit stream size is small. This allows sufficient time for water to infiltrate into the soil as it flows down the borders.

If the borders were too short for the stream size, the flow would reach the far end before sufficient water had infiltrated near the farm channel. This means that water must be left flowing to complete the irrigation and so some water would run-off the end of the border and be lost in the drainage system.

Borders can usually be much longer on the same soil when a larger unit stream size is available. This is because the water will spread more rapidly across the soil surface when the flow is increased.

Applying a larger irrigation depth also means that borders can be increased in length. A larger depth of water needs a longer contact time and so there is more time available to spread water along the border. This may not

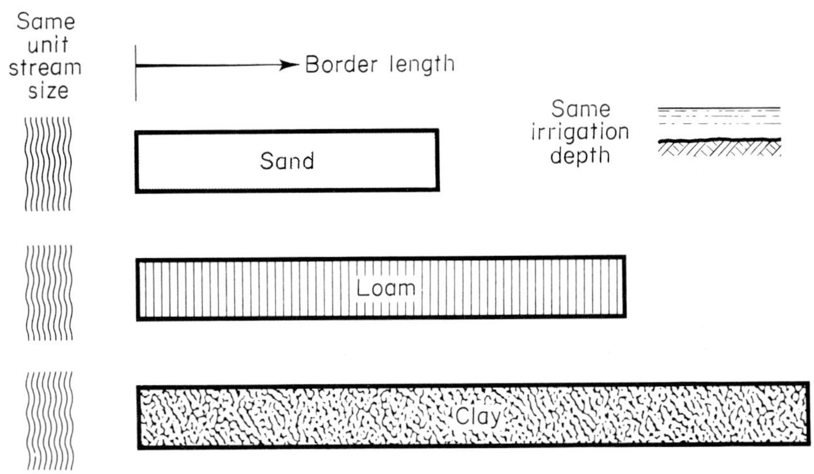

4.2(a)

be much of a problem on sandy soils, but on clays excessively long borders may be needed when large irrigation depths (100-200 mm) are applied. In such cases very low unit streams are used to slow down the advance and provide more time for infiltration.

On sloping land there is always the risk of soil erosion. On steep land slopes smaller unit stream sizes must be used to slow down the advance and so prevent erosion. This means that borders must be reduced in length as well if they are to be irrigated efficiently.

Border width largely depends on the length of border chosen and the available stream size. For example, if a 100 m long border on a sandy soil needs a unit stream of 10 ℓ/s and the available stream size is 50 ℓ/s, the border width is calculated as follows:

$$\text{Border width} = \frac{\text{available stream size } (\ell/\text{s})}{\text{unit stream size needed } (\ell/\text{s/m})}$$

$$= \frac{50}{10}$$

$$= 5\text{m}$$

Border width can also be affected by land slope. When there is a cross slope the land is usually levelled across each border and the 'staircase' effect is made in the same way that

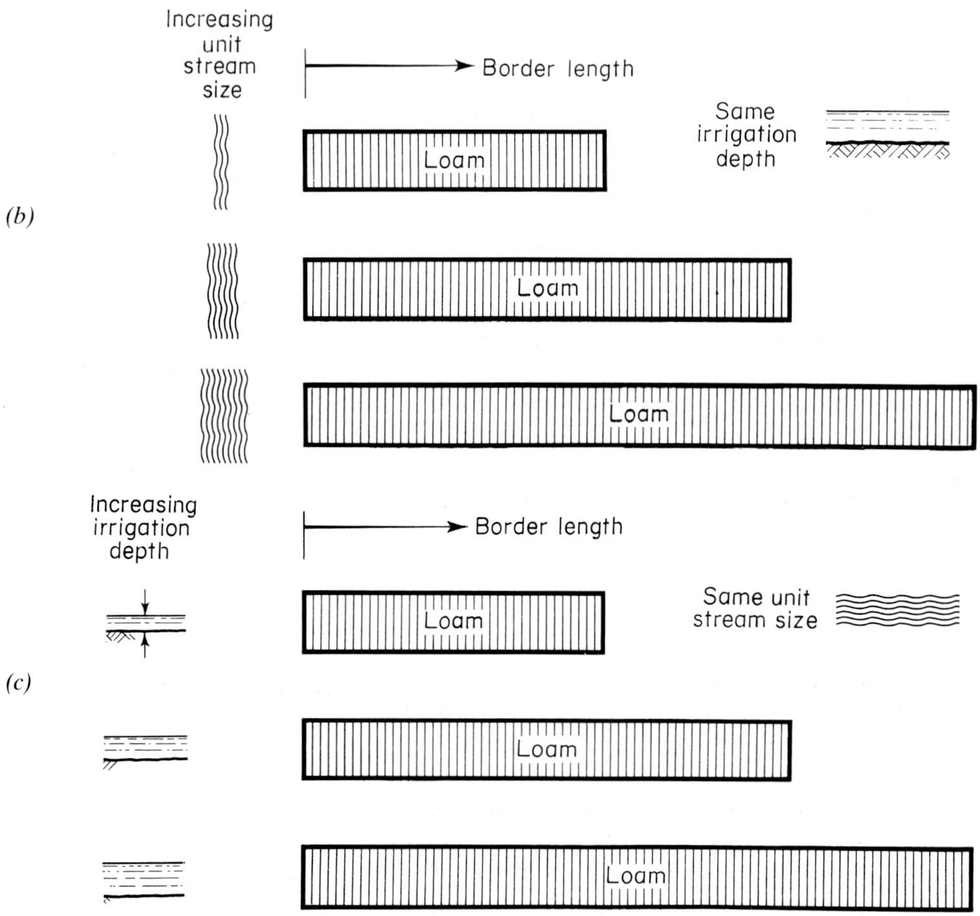

4.2 Factors affecting border size and shape. (a) Borders can be longer on clay soils than on sands (b) Borders can be longer when a larger stream size is used (c) Borders can be longer when a larger irrigation depth is applied (d) Borders are shorter on steeper land

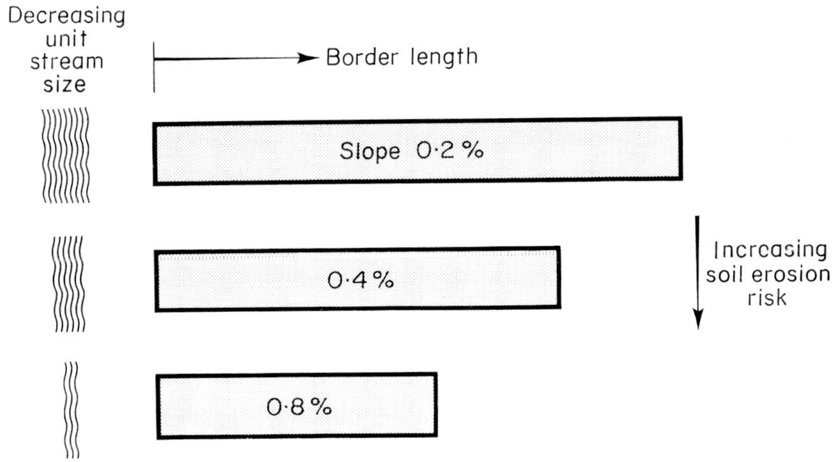

4.2(d)

level areas are made for basins on sloping land (section 3.2). The step between each border is usually limited to 60 mm to stop leakage through the earth bunds and this limits the width of the borders. For example, if there is a cross slope of 1% which must be levelled, a drop of 60 mm from one border to the next would limit the border width to 6 m.

As with basin irrigation, there are no simple calculations to help select the best dimensions for different soil types and unit stream sizes. It is usual to rely on local experience or the experience of others (table 4.1).

For example using table 4.1 when an irrigation depth of 100 mm is required on a sandy soil sloping at 0.2%, a border of length 60-100 m is needed with a unit stream size of 10-15 ℓ/s. On a clay soil the unit stream can be lower (3-6 ℓ/s) and the border needs to be much longer (350 m or more).

To summarise: (figure 4.2)

Borders can usually be much longer on clay soils than on sandy soils.

Borders can be longer when a larger unit stream size is available.

Borders can be longer when a larger irrigation depth is applied.

Borders must be shorter on steeper sloping land to prevent erosion.

Borders can be wider when larger stream sizes are available.

Table 4.1 Suggested border sizes

Soil type	*Irrigation Depth* (mm)	*Slope* (%)	*Width* (m)	*Length* (m)	*Unit Stream*[1] (ℓ/s)
Sand	100	0.2	12-30	60-100	10-15
		0.4	10-12	60-100	8-10
		0.8	5-10	75	5-7
Loam	150	0.2	15-30	90-300	4-6
		0.4	10-12	90-180	3-5
		0.8	5-10	90	2-4
Clay	200	0.2	15-30	350+	3-6
		0.4	10-12	180-300	2-4

[1] Unit stream is the flow for each metre width of border. The irrigation stream size is the unit stream multiplied by the width of the border (section 2.4)

Field shape and size

This may be a practical limit to the size and shape of borders. In small fields borders may be as long as the field or different lengths if the field is an irregular shape (figure 4.3). In large fields a common practice is to divide the land into two or more equal lengths. This makes it easier to deliver the same amount of water to each border.

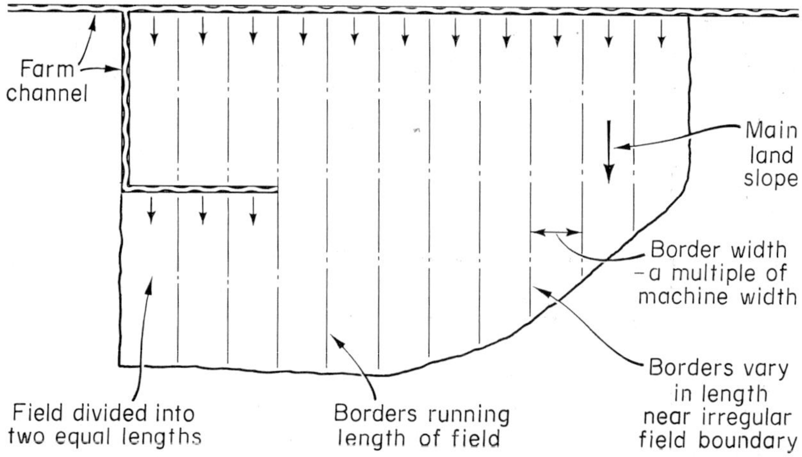

4.3 Effect of field size and shape on border size.

Farming practice

Border irrigation tends to be used mainly on the larger farms because of the long lengths needed for good water distribution. For this reason they are also very suitable for mechanised agriculture. Border widths should be some multiple of the width of the machine used. For example, if a cutter blade 2 m wide is used to harvest lucerne the border width sould be some multiple of 2 m.

4.3 Slope

Ideally borders should have a uniform slope along the length and no cross slope (figure 4.4). This is the slope across the width of the border.

The minimum slope is usually 0.1%. This ensures that water will flow down the border and any excess water can be easily drained.

The maximum slope depends on the risk of soil erosion. This depends on rainfall, soil

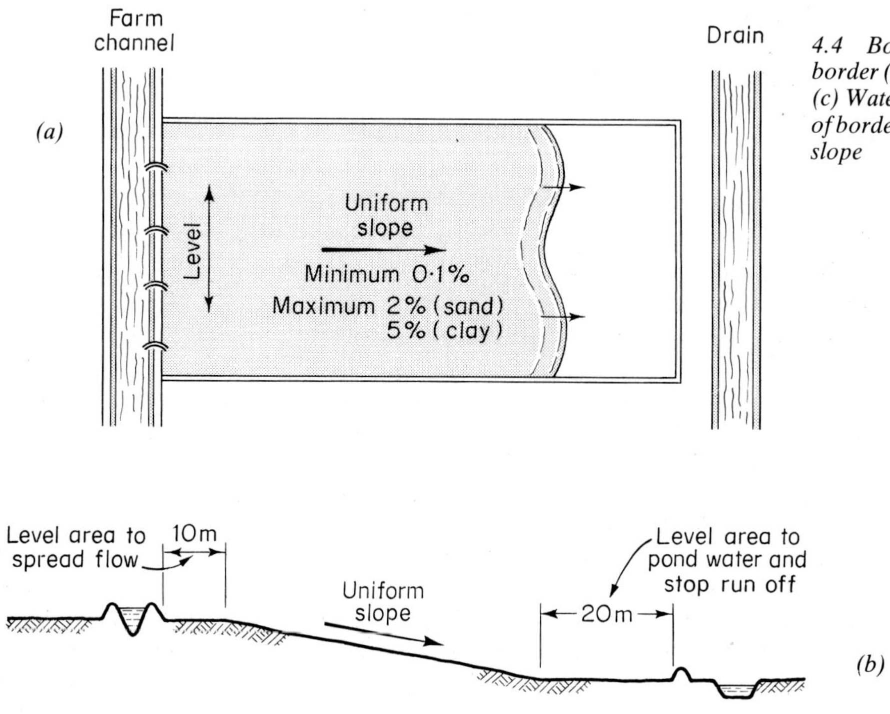

4.4 Border slopes. (a) Typical border (b) Section along border (c) Water spreads across width of border when there is no cross slope

4.4(c)

Table 4.2 Maximum land slopes for borders (%)

Soil type	Humid areas		Arid areas	
	Bare soil	Good crop cover	Bare soil	Good crop cover
Sand	0.3	1.0	1.0	2.0
Clay	0.5	2.0	2.0	5.0

type and crop cover (table 4.2). A good crop cover for reducing erosion is grass or lucerne.

When it is practicable, borders should be straight and aligned down the main field slope to eliminate any cross slope. This ensures that water flows evenly across the entire width of the border. Land slopes can be changed in a field by careful preparation but this work can be difficult and costly (chapter 7).

Some irrigators prefer the first 10 m along the border to be level as this helps to spread the water evenly across the border width. The last 20 m or so can also be levelled so that water can be ponded, as in basin irrigation, and run-off avoided.

WATER MANAGEMENT

4.4 Irrigating borders

When irrigating borders it is important to use the right unit stream size for the soil and land slope and to stop the flow at the right time so that just enough water infiltrates to fill the soil reservoir (section 2.11).

Deciding when to stop the flow can be a problem. If the flow is stopped too soon water may not even reach the end of the border. If it is left flowing too long, water will run off the end of the border. Irrigators often rely on general guidelines to decide when to stop the

(a)

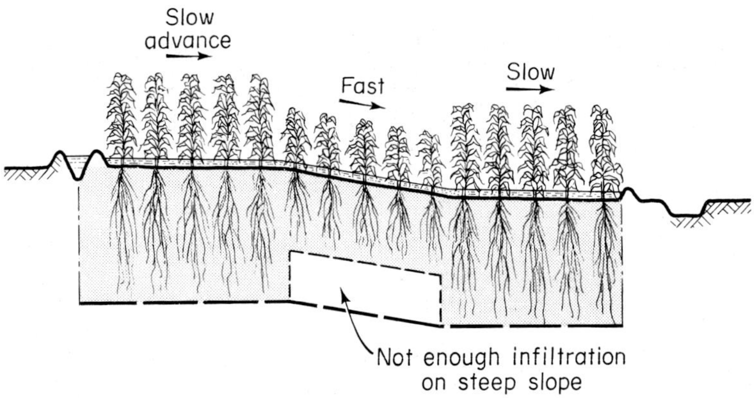

(b)

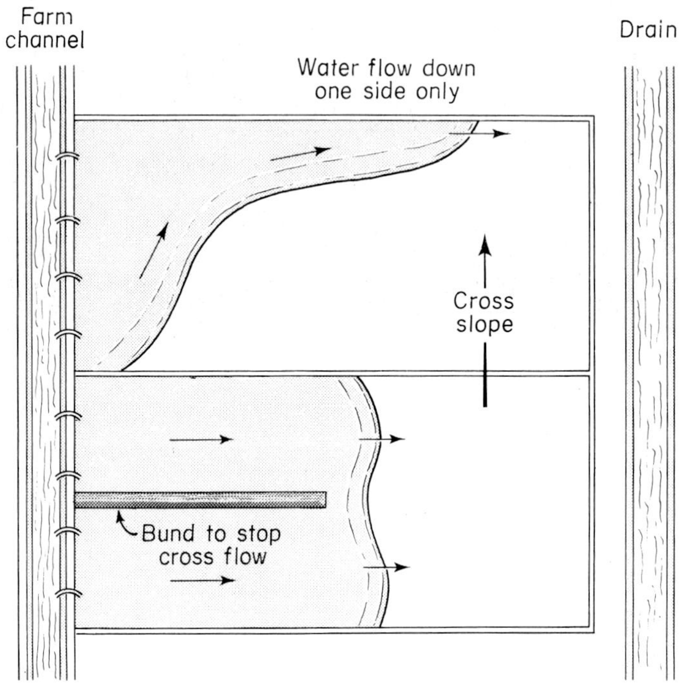

4.5 Common fault – poor land preparation. (a) Uneven border slopes (b) Effect of cross slope (c) Bund along border stops cross flow (d) Cross bund pushes water onto higher ground

flow. For example:

On clay soils the inflow is stopped when the advance reaches 0.6 of the way down the border.

On loamy soils inflow is stopped at 0.7-0.8 of the way down the border.

On sandy soils the advance must almost reach the end of the border before stopping the inflow.

However, these are only guidelines. Only practice in the field will really help the irrigator to make the right decisions.

Generally, on well managed border irrigation there should be very little deep percolation loss. Some run-off can be expected but this should be no more than 10-15% of the total amount of water applied. A small drain is needed at the end of each border (sometimes called a *tail drain*) to remove the excess surface water.

4.5(d)

4.5(c)

4.5 Common faults

Although border irrigation can be very efficient there are many common irrigation practices which result in poor water distribution and low irrigation efficiency. Some of these are described below.

Poor land preparation

When a border slopes unevenly the flow down the border is also uneven (figure 4.5). It flows slowly down the gentle slopes and faster down the steeper slopes. This affects the time available for infiltration and results in a poor distribution of water.

An uneven land surface having small depressions or high ground results in an uneven water distribution and water wastage in the same way as in basin irrigation, (section 3.11).

A cross slope in a border also causes problems. Water tends to flow down one side resulting in poor distribution across the border (figure 4.5). This can sometimes be corrected temporarily by constructing a bund

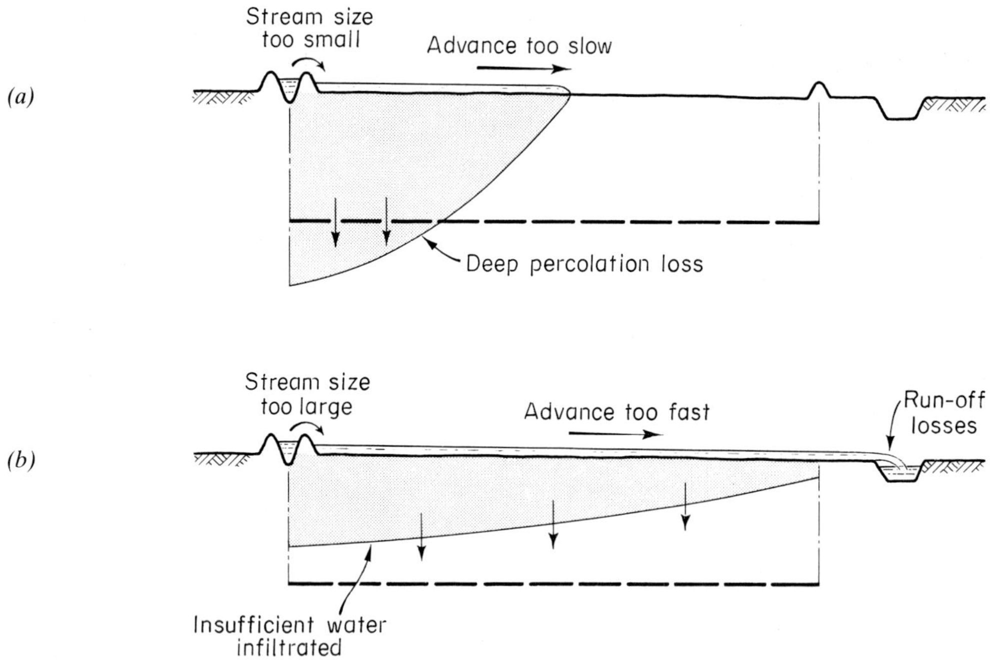

4.6 Common fault – using the wrong stream size.
(a) Stream size too small (b) Stream size too large

down the border to stop the cross-flow.

All these faults can be corrected by careful land preparation, (chapter 7).

Different soil types along border

The effect of this is very similar to that in basins, (section 3.11). Water distribution can be very uneven. The only solution is to realign the borders so that they contain only one soil type.

Using the wrong stream size

Once a border system has been installed the stream size is the only factor that can easily be changed.

Too much water applied usually means that the stream size is too small. The water takes a long time to flow down the border and reach some point when the inflow can be stopped. As a result too much water flows into the border and is lost as deep percolation (figure 4.6). A larger stream size would reduce these losses. The water would flow much faster and reach the point when the inflow can be stopped much sooner. As a result less water would flow into the border and the losses would be reduced. This problem usually occurs on sandy soils.

It seems strange at first to think that using a larger flow will reduce water losses. However the larger flow is used for a much shorter time. The discharge is higher but the volume of water required is much less.

Applying too little water usually means that the stream size is too large. The water flows too quickly down the border and reaches the point when the inflow should be stopped before sufficient water has time to infiltrate (figure 4.6). If the water is left flowing to give more time for infiltration then some will run off the end of the border. A smaller stream size would reduce these losses. Water would flow more slowly down the border giving more time for infiltration and reducing run-off. This is usually a problem on clay soils. A level area at the end of the border would allow water to be ponded and stop any run-off occurring.

To summarise:

Large stream sizes apply small irrigation depths.

Small stream sizes apply large irrigation depths.

Fixed irrigation schedule
Although the right stream sizes are used, some irrigators choose the irrigation time to fit in with other farm work rather than on the irrigation needed. Irrigation times to a fixed schedule such as 12 or 24 hours are common when only 6-10 hours are really needed. A lot of water can be wasted by this practice and fertilizers can also be washed out of the soil.

4.6 Efficiency
On well managed borders the irrigation efficiency can be as high as 80%. When borders are not used properly the efficiency can be much lower. The following are guidelines on the effects of poor practice on irrigation efficiency. The figures should be subtracted from 80%.

Common faults	*Subtract*
Poor land preparation	10-20%
Different soil types along border	5-10%
Using wrong stream size	10-15%
Fixed irrigation schedule	10-20%

5

FURROW IRRIGATION

5.1 Introduction

Furrow irrigation is the most widely used method for irrigating row crops. It is also the most misunderstood of all the irrigation methods. Water no longer flows over the entire soil surface but is confined to small channels (furrows) between the crop rows. It is gradually absorbed into the bottom and sides of the furrow to wet the soil (figure 5.1).

Furrow irrigation can be adapted to suit a

(a)

5.1 Furrow irrigation. (a) General view (b) Water flows in small channels

wide range of row crops, soils and farming practices. However, it is important to use the right shape of furrow, spacing and length and good water management for the method to work well.

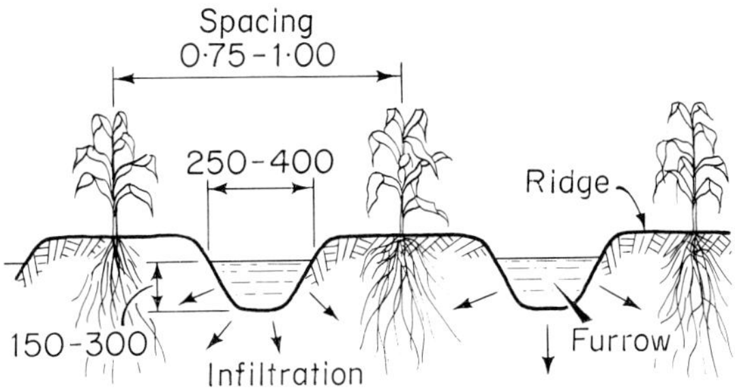

(b)

5.2 Shape

The shape of furrows is important to the adequacy and efficiency of furrow irrigation. Furrows are usually V-shaped. The width varies from 250-400 mm and the depth from 150-300 mm (figure 5.1). It depends on:–

Stream size

Soil type

Crops

Stream size
Each furrow is like a small channel and must be large enough to carry water without damaging the furrows. Furrow stream sizes are usually between 0.2 and 3.0 ℓ/s. The larger the stream size used, the larger must be the furrow channel.

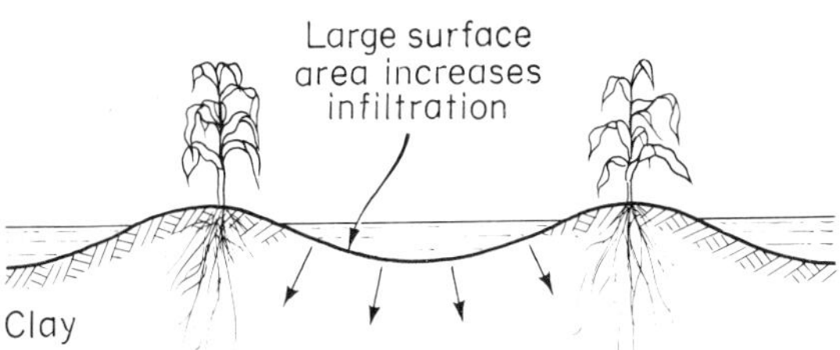

(a)

Soil type
On clay soils water infiltrates very slowly and so wide shallow furrows are sometimes needed to increase the soil area through which water can be absorbed more rapidly (figure 5.2). If the furrow is too narrow water will flow too quickly down the furrow before sufficient water has had time to infiltrate. In extreme cases the crop may not germinate near the top of the furrow because of a lack of water.

On sandy soils water infiltrates more rapidly than on clay soils and the problem is to try and reduce infiltration so that water will flow quickly along the furrow. For this reason narrow, deep furrows are used to reduce the soil area through which water can be absorbed.

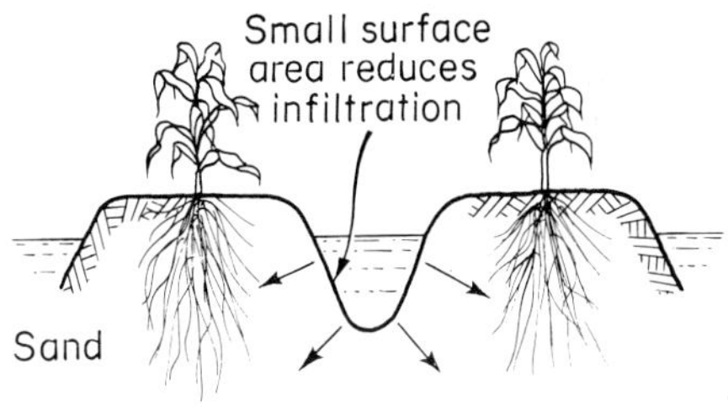

(b)

Crops
When seeds are sown or small plants are transplanted it is important that all the soil in the ridge is thoroughly wetted. This is best done by using shallow furrows. As the root system grows, furrows can be cut deeper to improve infiltration and increase the furrow discharge capacity.

For winter and early spring crops, the side slopes of ridges are sometimes changed to increase the soil exposure to the sun and so warm up the soil more quickly (figure 5.2).

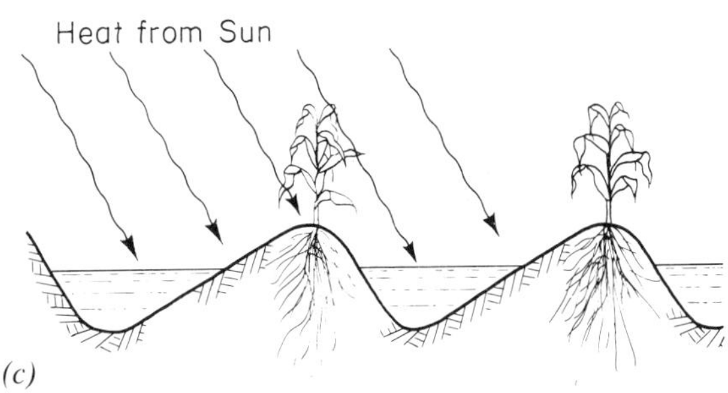

(c)

5.2 Furrow shapes. (a) On clay soil (b) On sandy soil (c) Side slope changed to warm up soil more quickly

5.3 Spacing

The Spacing between furrows depends on:–

Water movement in the soil

Crops

Cultivation practice

Water movement in the soil
This is the most important from an irrigation point of view. Water moves sideways (laterally) from furrows as well as infiltrating downwards. (figure 5.3). It will also move upwards to wet the tops of the ridges by a process called *capillary action.* Furrows must be spaced so that all the soil reservoir is properly wetted and this depends on the soil type.

On sandy soils the lateral movement of water is usually small and so furrows need to be close together (0.5 m). On clay soils there is much more lateral wetting and furrow spacing can be 1.2 m or more. If the furrow spacing is too wide there will be a dry area in between the furrows and the crop may not get enough water.

Crop
Row crops are normally spaced 0.75-1.0 m apart for the convenience of planting, cultivating and harvesting. Some crops, such as vegetables can be planted in double rows. The ridge is then made much wider than for single planting and is often called a *bed.*

Cultivation practice
On many farms the equipment available for cultivation and harvesting may also determine furrow spacing. The same row widths are often chosen for different crops and soils so that the same implements can be used throughout. However, it is important that the width chosen will ensure sufficient wetting of all the soils on the farm.

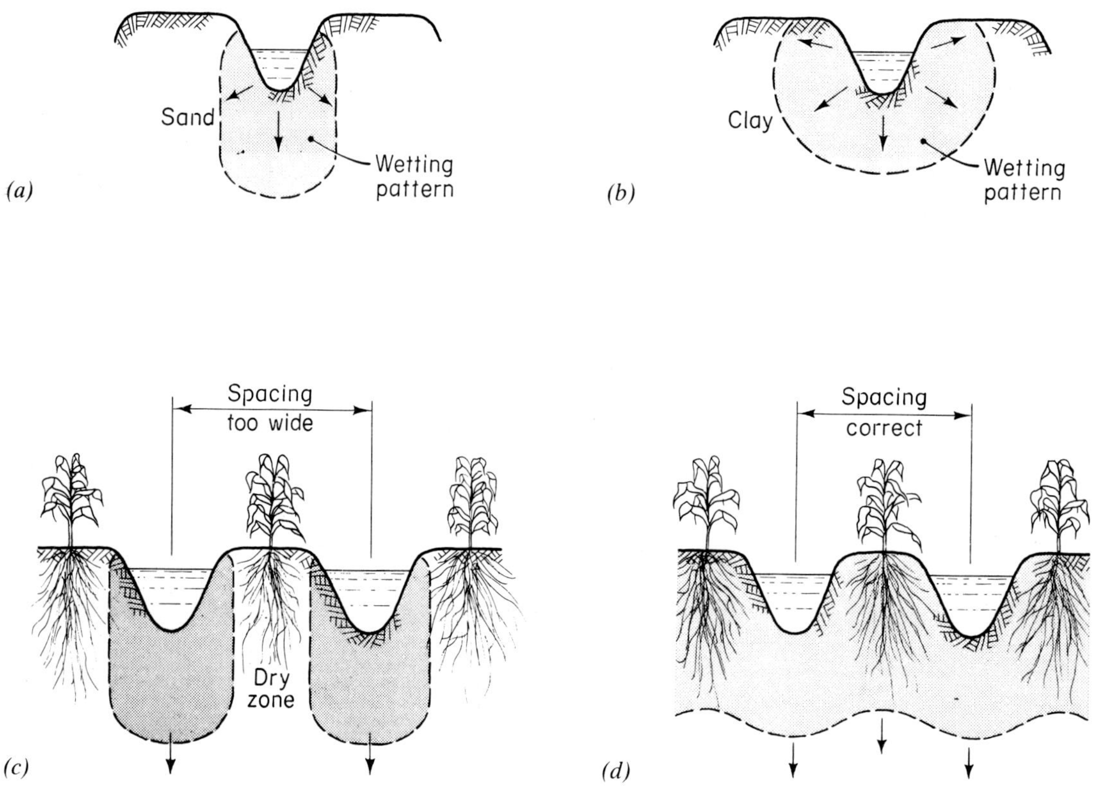

5.3 *Furrow spacing. (a) Sandy soil (b) Clay soil (c) Spacing too wide (d) Right spacing*

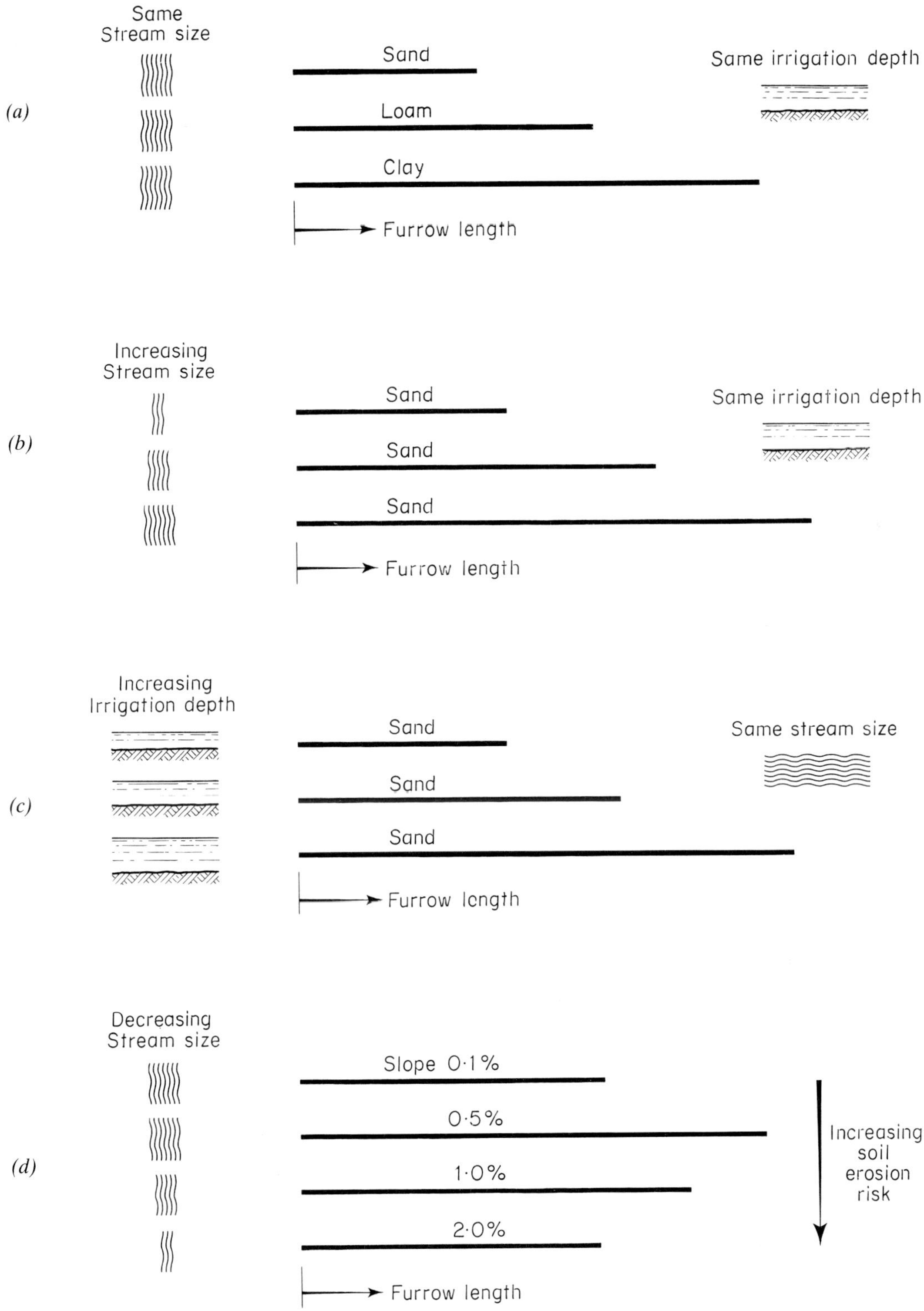

5.4 Factors affecting furrow length. (a) Furrows can be longer on clay soils than on sands (b) Furrows can be longer when larger stream sizes are used (c) Furrows can be longer when larger irrigation depths are applied (d) Furrows are usually shorter on steeper sloping land to prevent erosion

5.4 Length

The choice of furrow length depends on the following factors::–

Soil type

Stream size

Irrigation depth

Field size and shape

Slope

Farming practice

Soil type, stream size, irrigation depth and slope

These factors are the most important in determining what length furrows can be (figure 5.4). To understand this it is useful to remember the way in which furrows are normally irrigated (section 2.11).

When irrigating sandy soils water infiltrates rapidly. This means that furrows must be short (100 m or less) for water to flow quickly along to the far end even when quite large stream sizes are available. If the furrow was too long for the stream size, too much water would be lost in deep percolation near the farm channel.

On clay soils water infiltrates more slowly and so furrows can be longer than on sands (up to 800 m or more) even when only small stream sizes are available. If the furrows are not long enough for the stream size, excessive run-off can occur. This is because water cannot be ponded in furrows as it can in basins and so water must be left flowing until sufficient has been absorbed into the soil.

Furrows can be longer on the same soil when a larger stream size is available. This is because water will advance more rapidly down the furrows when the flow is increased. However, there is a limit to the size of stream that can be used because of the risk of soil erosion. (max 3.0 ℓ/s).

Applying larger irrigation depths also means that furrows can be longer because there is more time available to spread the flow down the furrows.

When furrows are used on steeper sloping land (up to 0.3%) they can usually be longer. This is because water moves more rapidly down the furrow as the slope increases. However on sloping land there is always the risk of soil erosion. As the land becomes steeper than 0.3%, smaller stream sizes are used to prevent erosion occurring. This means that shorter furrow lengths must be used.

As in the case of basins and borders there are no simple ways to calculate the best furrow length for different conditions. It is again usual to rely on local experience or the experience of others (table 5.1).

For example using table 5.1 on a clay soil with a slope of 0.1% the maximum stream size is 3.0 ℓ/s and the best length of furrow is 340 m for an irrigaiton depth of 75 mm. If the application depth was increased to 150 mm the furrow length could also be increased to 440 m.

Table 5.1 Suggested furrow lengths (m)

Slope (%)	*Maximum stream size* ℓ/s	*Clay* 75	150	*Loam* 50	100	150	*Sand* 50	75	100
		Average irrigation depth (mm)							
0.05	3.0	300	400	120	270	400	60	90	150
0.1	3.0	340	440	180	340	440	90	120	190
0.2	2.5	370	470	220	370	470	120	190	250
0.3	2.0	400	500	280	400	500	150	220	280
0.5	1.2	400	500	280	370	470	120	190	250
1.0	0.6	280	400	250	300	370	90	150	220
1.5	0.5	250	340	220	280	340	80	120	190
2.0	0.3	220	270	180	250	300	60	90	150

5.5 *Water advances faster along furrows compacted by tractor wheels.*

To summarise: (figure 5.4).

Furrows can usually be longer on clay soils than on sandy soils.

Furrows can be longer when a larger stream size is used for irrigation.

Furrows can be longer when a larger irrigation depth is applied.

Furrows are usually shorter on steeper sloping land to prevent erosion.

Field size and shape

This may be a practical limit to the length of furrows. In small fields furrows may be as long as the field. Larger fields may be divided into two or more equal lengths as in the case of borders. This makes it easier to deliver the same amount of water to each furrow.

Farming practice

From a farming point of view furrows should be as long as possible. Fewer farm channels and drains are needed, less land is taken up by the channels and it is easier to mechanise. Short furrows require a lot of attention because the flow must be changed frequently from one furrow to the next.

The use of tractors on the farm can also affect furrow lengths. Furrows are often compacted by tractors tyres and this can reduce soil infiltration rates. This can be of benefit on sandy soils as furrow lengths can be increased without extra percolation loss. However, all the furrows would need similar compaction otherwise water would advance at different rates (figure 5.5).

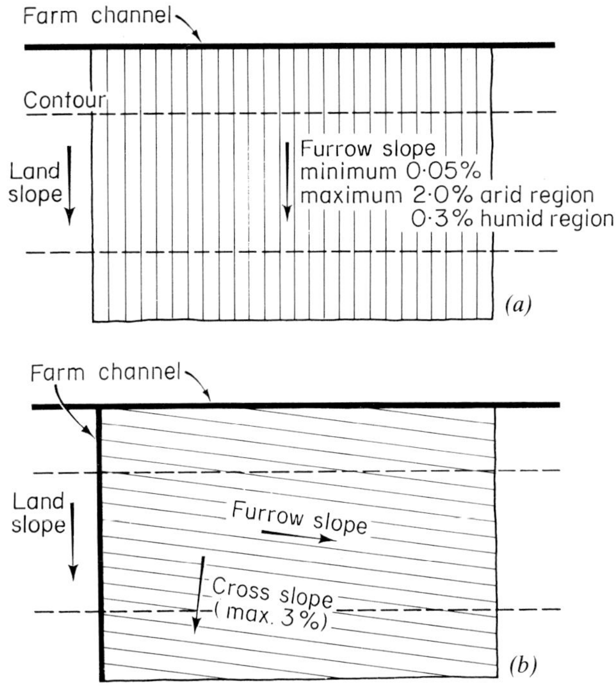

5.6 *Furrow alignment on sloping land. (a) Land slope less than maximum furrow slope (b) Land slope greater than maximum furrow slope (c) Erosion risk when cross slope is too steep (d) Furrows on undulating land*

5.5 Slope

Ideally furrows should have a uniform slope. A minimum slope of 0.05% is needed to ensure that water will flow down the furrow and any excess water can be drained.

The maximum slope depends on the risk of soil erosion. This is more critical in furrows than in borders because the flow is confined in a small channel and so it can erode the soil more easily. In arid regions the maximum furrow slope is 2.0%. In more humid regions where there is a risk of intensive rainfall the maximum slope is 0.3%.

The maximum allowable slope is usually related to a non-erosive stream size (table 5.1). This is the maximum stream size that will not cause any erosion in the furrow. Furrows on slopes of 0.05% and 0.1% could safely carry flows greater than 3 ℓ/s from an erosion point of view. In practice however, this would be beyond the capacity of most furrows and overtopping would occur.

When it is practicable, furrows should be

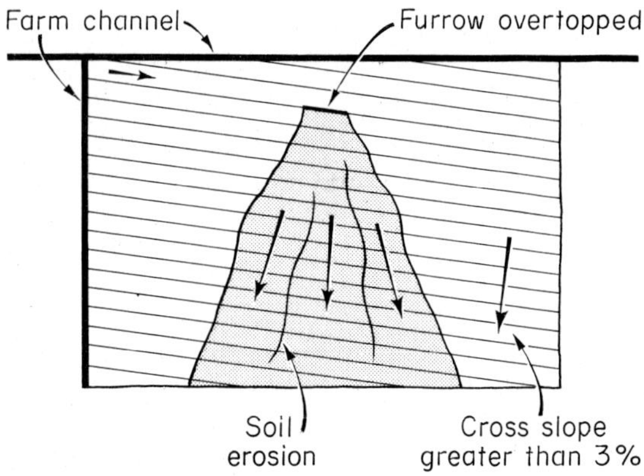

5.6(c)

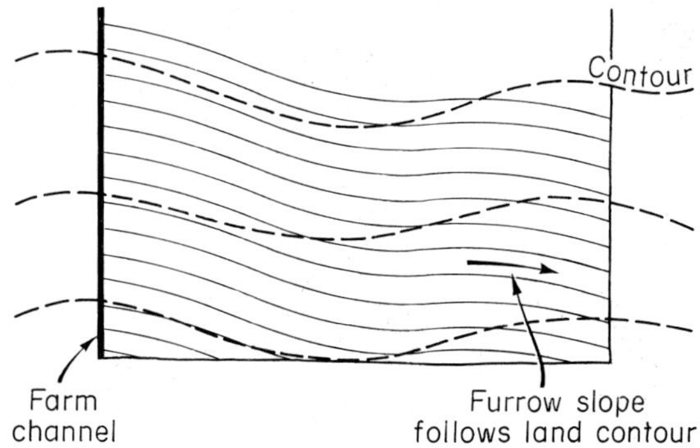

5.6(d)

straight and parallel to the edge of the field and aligned down the main land slope. If the land slope is too steep furrows can be aligned across the main slope to reduce the furrow slope (figure 5.6). However the land slope should be less than 3% to avoid any risk of serious soil erosion. If the cross slope was greater than this and a furrow near the top of the field overtopped it could seriously damage the field and wash away the topsoil. Great skill is needed to irrigate in these conditions.

On undulating land furrows are sometimes set out along the land contours. They are called *contour furrows*. Cultivating and irrigating furrows in this way again requires great skill. This should not be done on sandy soils or on clay soils which tend to crack.

Some furrows are used on level ground but this is usually when irrigating heavy clay soils with low infiltration rates.

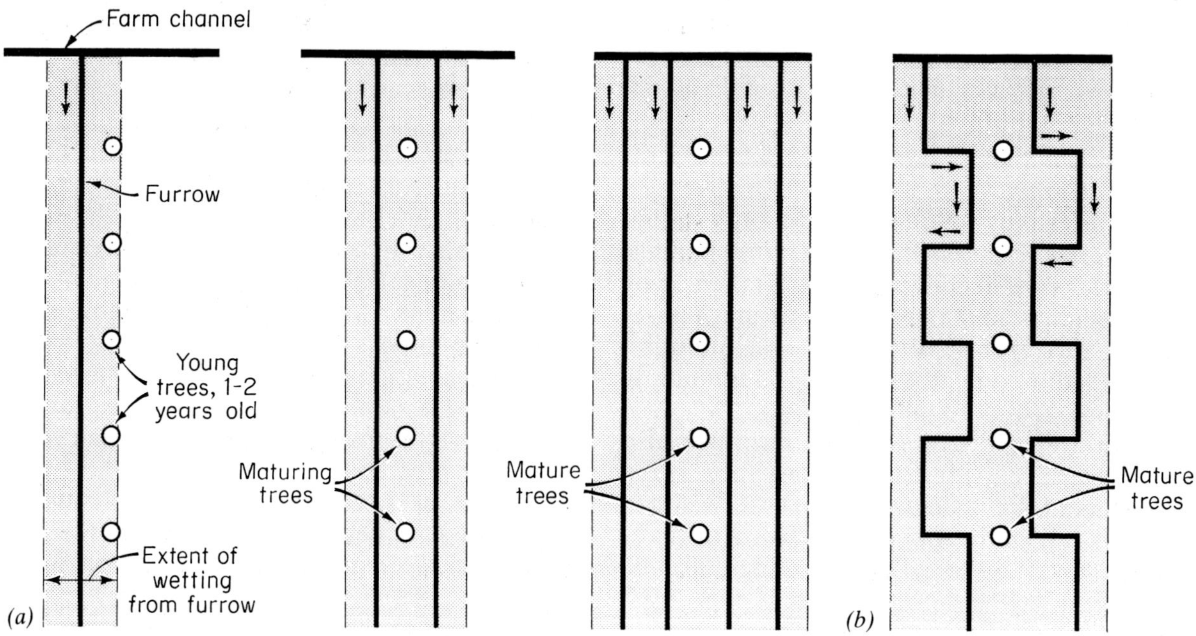

5.7 Using furrow irrigation in orchards. (a) Number of furrows increased as trees grow larger (b) Zigzag furrows used to improve spread of water (c) Furrows irrigating an orchard

5.7(c)

5.6 Crops

Most row crops such as vegetables, cotton, sugar beet and potatoes are irrigated by furrow methods. The crops are usually planted on raised beds or ridges with furrows spaced between them.

Orchards and vineyards can also be irrigated with furrows. When the trees are young a single furrow may be enough to wet the root zone. As they grow, more furrows can be placed down between the tree rows (figure 5.7). Sometimes special furrow layouts are used such as zig-zag patterns to improve the spread of water in the soil. An advantage of furrows is that not all the soil surface is wetted and so there is good access in the orchard even during irrigation.

When saline water is used for irrigation crops which are sensitive to salinity are usually planted on the side of the ridge and not on the top (figure 5.8). This is because salt which is not leached down into the soil usually accumulates on the ridge top. In arid regions where irrigation water is limited plants are often grown in the furrow. This helps plants to obtain as much water as possible and keeps them well away from any salinity.

WATER MANAGEMENT

5.7 Irrigating furrows

When irrigating furrows, water is supplied to each furrow from the farm channel using siphons or spiles. Several furrows can be irrigated at the same time depending on the discharge available in the farm channel. The stream size must be large enough for the flow

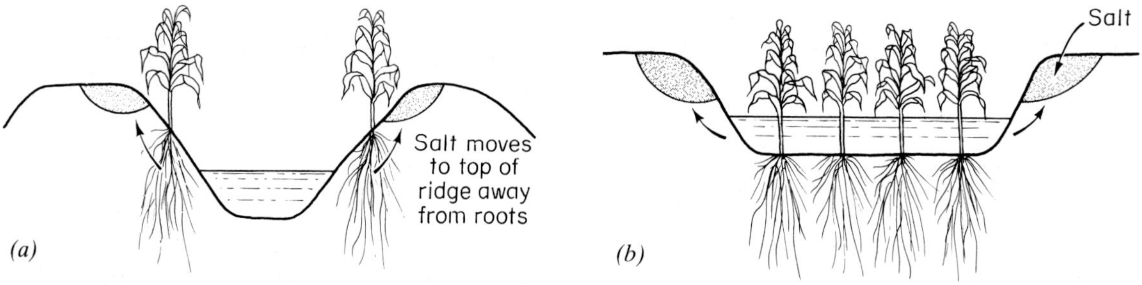

5.8 Using furrow irrigation with saline water. (a) Crop planted on side of furrow (b) Crop planted in bed of furrow

to advance quickly down the furrow. This will ensure a reasonably uniform irrigation (section 2.11).

Some water will always be lost in deep percolation but this will be quite small if the right stream size is used.

Run-off can be a major problem. Even on well-managed furrows, run-off can be as much as 30% of the inflow. For this reason it is important to provide a shallow drain at the end of the field (sometimes called a *tail drain)* to remove this excess water. On many farms this is not fully understood and drains are not provided. This can result in water logging which may damage sensitive crops (figure 5.9). There are several ways in which irrigators can reduce run-off losses.

One way is to allow some water to pond at the end of a furrow and to back up into adjacent furrows. Any water which does not infiltrate within 24 hours must be drained before any damage is done to the crop.

Another way is to reduce the stream size and relax the quarter time rule. This is a common practice on clay soils where deep percolation losses will be small even if the advance only reaches the end of the furrow in half the contact time (T/2).

A third method is to reduce the stream size once the flow has reached the end of the furrow. This is called the *cut-back* method. The stream size is usually cut back (or reduced) once or twice during an irrigation.

On some well organised schemes run-off water is collected in storage ponds and pumped through a pipeline back into the irrigation system (figure 5.9). This must only

5.9(a)

be done with great caution because run-off water may become mixed with saline drainage water and this can result in damage to crops. As a general rule drainage water is best left alone unless the irigator is sure it is safe to use.

All of these practices can greatly improve furrow irrigation efficiency but they can be very difficult to put into practice. Usually irrigators have enough problems just trying to irrigate without the extra problem of coping with run-off. A great deal also depends on the irrigation system and on the skill of the irrigator for these practices to work well.

(b)

5.9 *Dealing with run-off from furrow irrigation.*
(a) Water logging – no drainage provided (b) Releasing water into a tail drain (c) A return-flow system

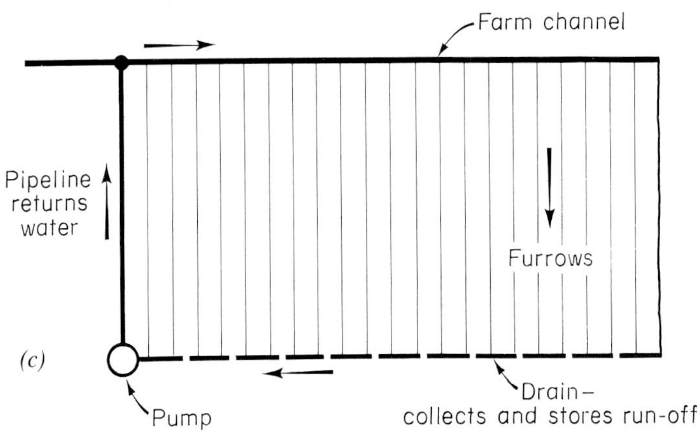

(c)

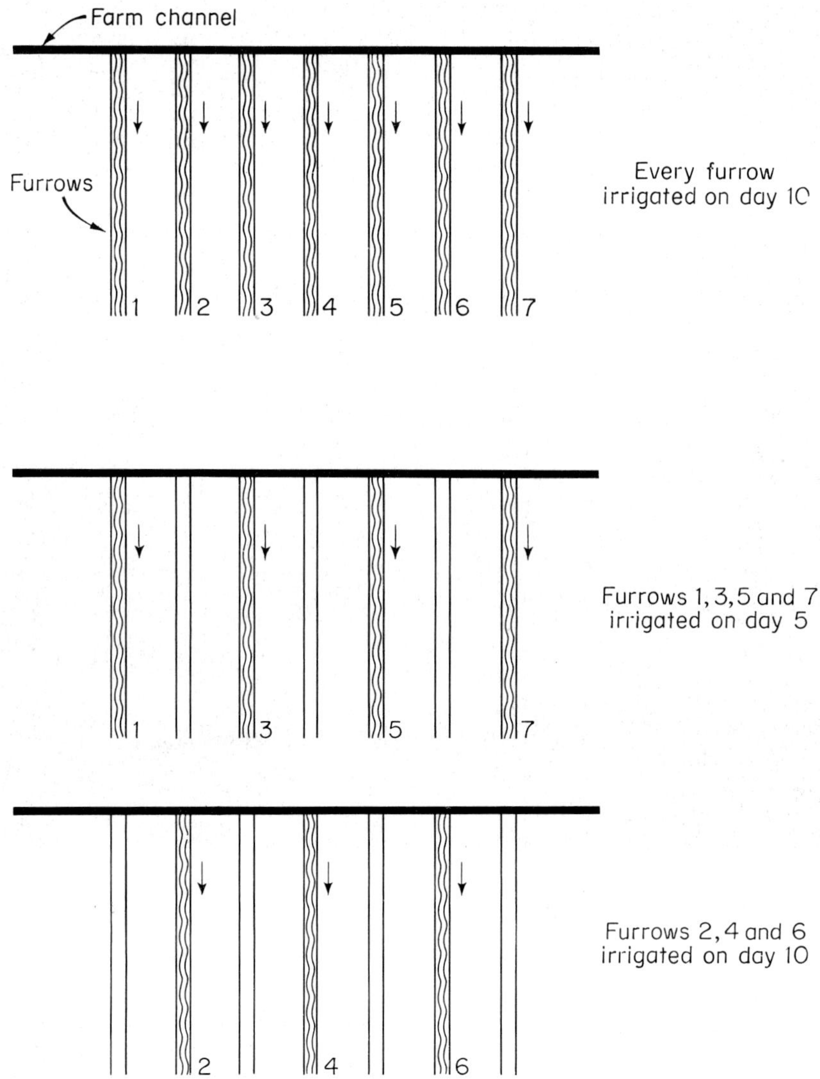

5.10 Alternate furrow irrigation.

5.8 Alternate furrow irrigation

This method is very useful when there is a water shortage. A limited amount of water can be spread over a large area so that all the crops obtain some water.

The practice involves irrigating every second (alternate) furrow in a field instead of irrigating every furrow (figure 5.10). For example, if a crop is normally irrigated every 10 days, alternate furrow irrigation would mean irrigating furrows 1, 3, 5 and 7 on day 5 and furrows 2, 4 and 6 on day 10. The crop receives some water every 5 days instead of a large amount every 10 days. This is usually better for the crop.

When there is a shortage of water the normal interval between irrigations may be increased from 10 to 14 days and this can reduce the crop yield. By using the alternate furrow method all the crop would receive some water every 7 days.

This method is often used to irrigate orchards but water is usually applied to alternate tree rows rather than to alternate furrows. This is called *alternate middle irrigation.* It means that there is always reasonably good dry access into the orchard.

5.11 Poor land preparation.

5.12 Common fault – poor crop growth at head of furrows on steep clay soils.

5.9 Common faults

Although furrow irrigation can be very efficient there are many common irrigation practices which result in poor water distribution and low irrigation efficiency. Some of these are described below.

Poor land preparation

This is very similar to the problem described for border irrigation. Uneven slopes and low areas in a field can result in poor water distribution and water logging (figure 5.11). Very small changes in ground level can have significant effects on irrigation efficiency.

All these faults can be corrected by careful land grading. See chapter 7.

Furrows on clay soils
On clays infiltration is usually very slow and sometimes there is difficulty in getting enough water to infiltrate into the soil. On sloping land the water may flow too quickly down the furrow with the result that only a small amount of water infiltrates at the head of the furrow causing poor crop growth (figure 5.12). This problem can be overcome by using a very small stream size or regrading the land so that the slopes are much flatter. In either case the advance will be slower and more water will infiltrate at the head of the field.

Different soil types along furrow
The effect of this is very similar to that for basins and borders. Water distribution will be very uneven. The only solution is to re-align the furrows so that they contain only one soil type.

Advance time too long
This is usually the result of using a stream size which is too small. It is done for two reasons. Firstly, the irrigation may be to a fixed schedule such as 8 or 12 hours to fit in with other farm work. Secondly, it reduces run-off because water is slow in reaching the end of the furrows (figure 5.13). However, this practice greatly increases deep percolation losses. These losses are not so obvious as run-off which can be easily seen. Only by choosing the right stream size for the soil conditions can losses be reduced.

Stopping inflow too soon
Even when the right stream size is used, the inflow is sometimes stopped too soon to try and reduce run-off. This results in a poor distribution of water and plants at the ends of furrows do not get enough water (figure 5.14).

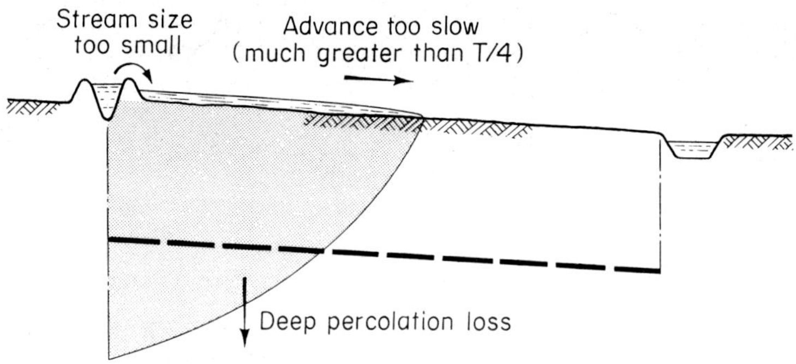

5.13 Common fault – advance time too long.

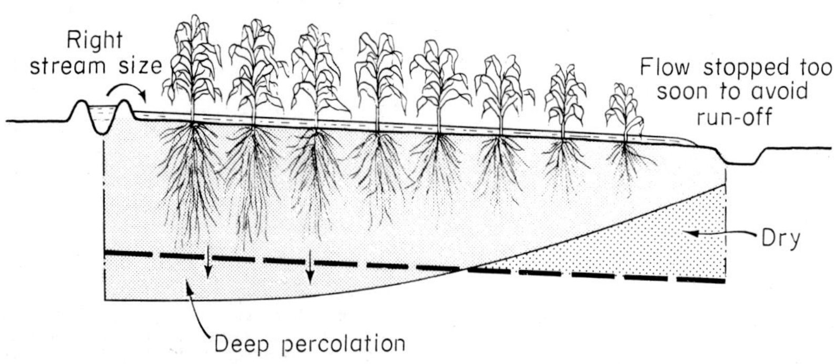

5.14 Common fault – stopping inflow too soon.

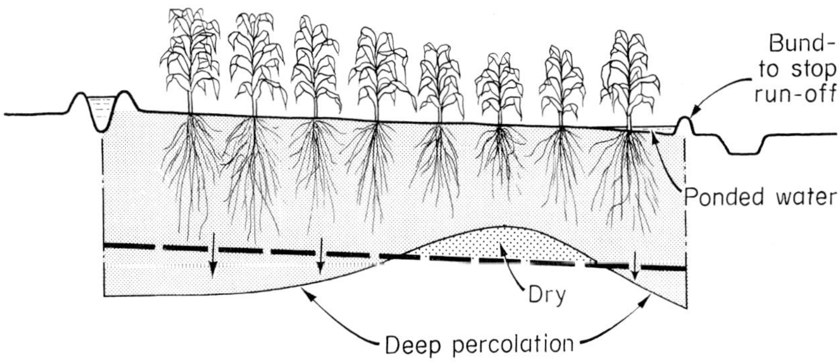

5.15 Common fault – bunding end of furrows.

This can be seen very clearly in some crops. The growth is much poorer towards the end of the furrows.

Bunding end of furrows

To try and correct the above fault some irrigators bund the ends of furrows to stop all run-off (figure 5.15). This improves the infiltration at the far end because water is ponded but there is still not enough irrigation further up the slope. This dry area is often called a *hot-spot* and is clearly visible in some crops because of the poor growth.

5.10 Efficiency

On well managed furrows the irrigation efficiency can be as high as 90%. When furrows are not used properly the efficiency can be much lower. The following are guidelines on the effects of poor practice on irrigation efficiency. The figures should be subtracted from 90%.

Common faults	*Subtract*
No return flow system	20-40%
Poor land preparation	10-20%
Different soil types along furrow	5-10%
Advance time too long	10-20%
Stopping inflow too soon	10-20%

6

CHOOSING AN IRRIGATION METHOD

6.1 Introduction

Many factors must be considered when deciding which is the best method of surface irrigation to use (table 6.1). These include:

- Land topography
- Soils
- Field shape
- Crops
- Labour

6.2 Land topography

The slope of the land is most critical when selecting an irrigation method. If the land is level or less than 0.1% basin irrigation can be used with very little land preparation. If the land slope is more than 0.1% it may be better to use border or furrow methods. Basin irrigation could be used but carefully constructed level terraces would be needed.

The maximum slope depends on the risk of soil erosion. The slope can be much greater when good ground cover crops such as grass and lucerne are grown because there is much less risk of erosion. Also in humid areas maximum slopes are less because of the risk of soil erosion from high rainfall. Note that it is not usual to plant ground cover crops on furrows and so only slope values for bare soil are given in table 6.1.

When the land is undulating or uneven it can be re-graded to a level or uniformly sloping surface to suit a particular irrigation method.

6.3 Soils

All the surface methods are suitable for soils of low (1-10 mm/h) and medium (10-30 mm/h) infiltration rate soils. If infiltration rates are greater than 30 mm/h it may be difficult to get a high enough stream size to flow a reasonable distance across the soil before it is all absorbed. In such cases sprinkler or trickle irrigation should be used.

6.4 Field shape

Basin irrigation is easily adapted to irregular shaped fields. Borders and furrows are best suited to rectangular fields so that their lengths are the same. This makes them easier to irrigate because they will all need the same stream size and irrigation time.

6.5 Crops

Crops are grouped into four main categories (table 6.1) — row crops, close growing crops which are sown or drilled, flooded rice and orchards.

All surface methods can be used to grow row crops but furrows and furrowed-basins are most suited to crops that do not like to stand in water for 12 hours or more.

Close growing crops can be grown in basins and borders but not on furrows.

Flooded rice needs level land on which to pond water and so is only suited to basins.

Orchards can be irrigated using any of the methods which are suited to the field shape and land slopes.

6.6 Labour

Labour is needed to use all the irrigation methods. The cost of hiring the labour, its availability and the skills available may all affect the choice of method.

Basin irrigation requires the least amount of labour. Borders and furrows usually need more labour and greater skill to irrigate efficiently.

When irrigating for the first time it is important to choose a simple method such as basin irrigation. When labour gains experience the more complex methods of borders and furrows can be used.

Table 6.1 Factors affecting choice of surface method

IRRIGATION METHOD	*LAND TOPOGRAPHY* Humid regions *Bare soil*	Humid regions *Good crop cover*	Arid regions *Bare soil*	Arid regions *Good crop cover*	*SOIL INFILTRATION RATE* (mm/h)	*FIELD SHAPE*	*CROPS* *Row crops*	*Sown, drilled crops*	*Flooded Rice*	*Orchards*	*LABOUR* *(hours/ha. per irrigation)*
Basin	Level slope or less than 0.1 (steeper slopes need terracing)				up to 30	Any shape	Yes	Yes	Yes	Yes	0.5 - 1.5
Border	0.5	2.0	2.0	5.0	up to 30	Rectangular	Yes	Yes	No	Yes	1.0 - 3.0
Furrow	0.3	–	2.0	–	up to 30	Rows should be equal length	Yes	No	No	Yes	2.0 - 4.0
	cross slope 3.0		cross slope 3.0								

7

LAND PREPARATION

7.1 Introduction

Good land preparation is most important for efficient surface irrigation. It involves excavating (cutting) soil from high ground and removing it to fill in hollows to form a level or uniform sloping land surface (figure 7.1). In most cases only small depths of soil are moved (200 mm or less) but the work must be done very accurately. Some of the effects of poor land preparation on basin border and furrow irrigation have already been described. Even small differences in land level can cause poor water distribution and low irrigation efficiency.

Good land preparation will help to:

Increase irrigation efficiency

Improve water distribution

Reduce soil erosion

Reduce water logging

Land preparation is often called *land levelling* but since the land can also be prepared to a slope for border and furrows it is sometimes called *land grading.*

7.2 Factors affecting land preparation

Making a plan for land preparation can be a complex job and usually requires the services of an experienced engineer and soil scientist. They must decide on the size of land area to be prepared and the land slope (gradient) needed so that:

The amount of soil excavated (cut) is equal to the fill required.

The minimum amount of soil is moved.

This means carefully examining:

Soils

Topography

Irrigation method

Rainfall

Farming practices

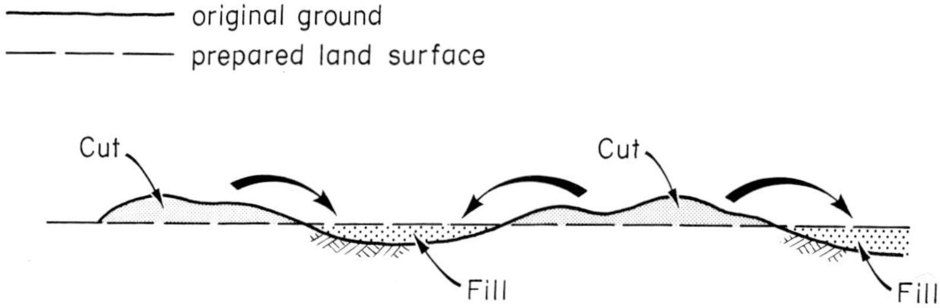

7.1 Land preparation. 1. Soil cut is equal to the fill required 2. Minimum amount of soil is moved

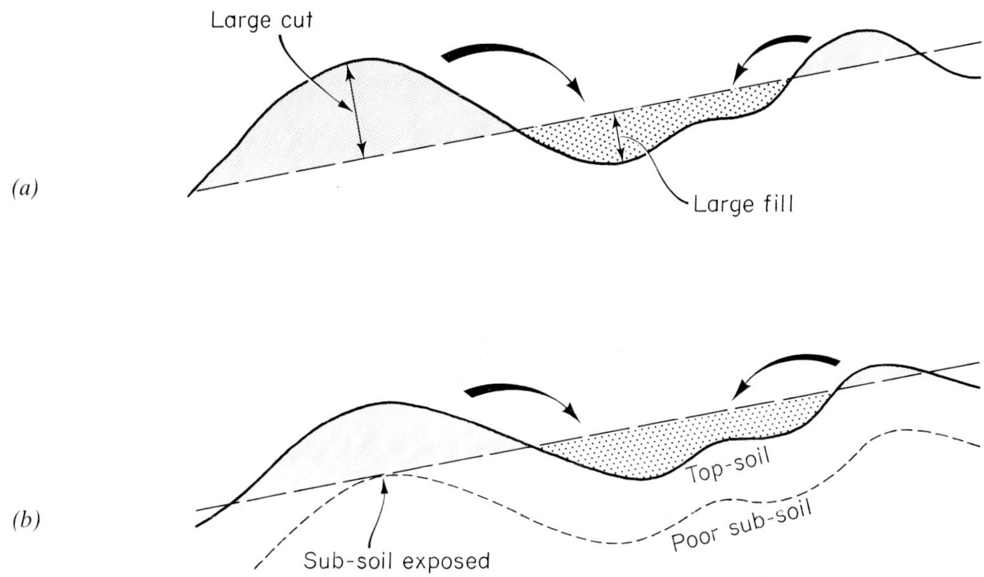

7.2 Effect of soil on land preparation. (a) Deep fertile soil (b) Shallow top soil

Soils

A survey of the soil is usually the first step in deciding just how much soil can be safely excavated. On very deep fertile soils quite large cuts can be made to obtain a uniform grade (figure 7.2). On shallow soils even small cuts may uncover saline soils, dense, compact soil layers or gravel which may be infertile and so reduce crop yields. Sometimes it is possible to remove all the shallow top soil completely, grade the sub-soil and then replace the top soil, but this can be a difficult and expensive operation. An alternative is to use a method of irrigation which does not require any preparation such as sprinkler or trickle irrigation.

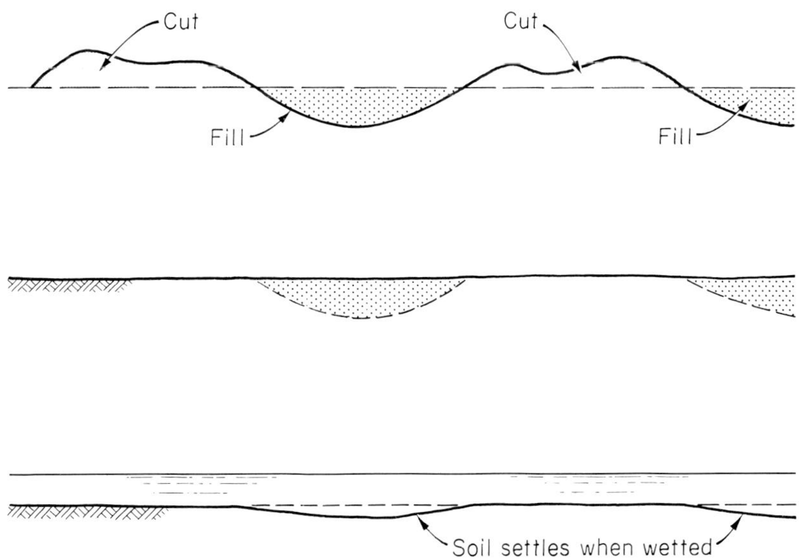

7.3 Soil settlement problems.

7.4 *Using profile boards to check land gradients.*

Topography

A survey of the land surface is also needed so that the engineer can locate the high and low spots and plan the earth movement. He must try to balance the amount of excavation and fill required to give the least amount of earth movement. In practice soil tends to settle in the hollows when it is wetted and so the volume excavated needs to be greater than the volume of fill by about 10-20% (figure 7.3).

The land usually must be surveyed again after preparation to check that the work has been done properly. This can be very time consuming work. Profile boards are often used as a simple and quick method of checking (figure 7.4).

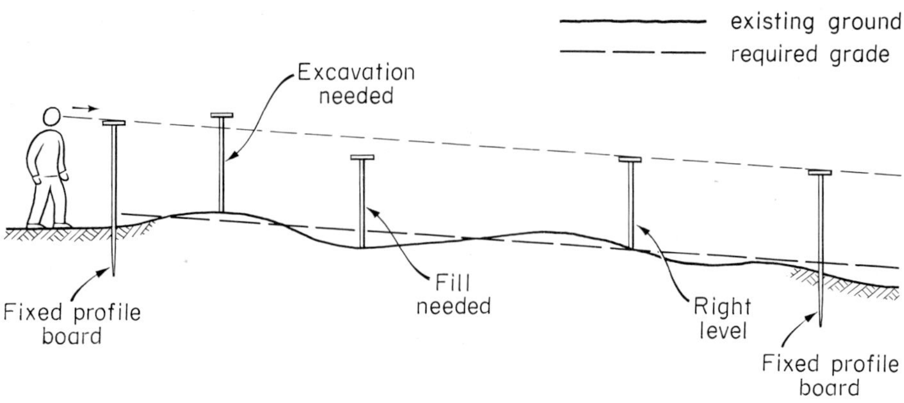

7.4 *Using profile boards to check land gradients.*

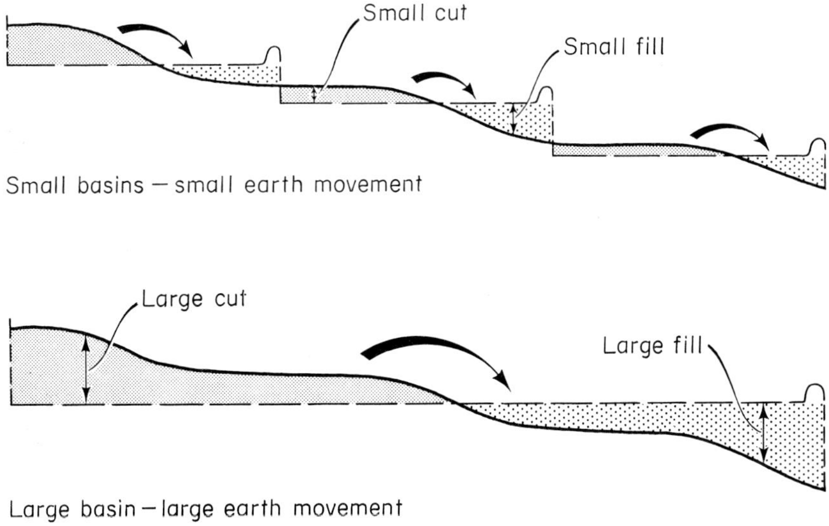

7.5 Effect of basin size on earth moving.

Irrigation method
This determines the slope to which the land is graded. For basin irrigation the land surface must be level. For border irrigation the main slope should be between 0.1-5% depending on soil type and crop cover and there should be no cross slope. For furrow irrigation the main land slope should be between 0.05-2%.

The slope of the finished surface must fall within these limits but be as near as possible to the average land slope. This will keep earth moving and costs low.

Irrigation method will also help to decide just how large an area to prepare to the same gradient. In basin irrigation, for example, each basin can be levelled separately. If basins are small then the area to be levelled will be small. If larger basins are used then a larger area must be prepared to the same level. Generally, the larger the area to be levelled, the greater will be the amount of earth moving needed and the greater the cost (figure 7.5). On large mechanised farms this extra cost may be justified because it will provide large areas in which machines can work easily.

Rainfall
In areas of high rainfall the choice of land slopes can be limited by the risk of soil erosion and the need for good drainage. For furrow irrigation the maximum slope should not exceed 0.3%. Borders can be much steeper (up to 2%) depending on soil type and crop cover. The smallest slopes on furrows and borders is usually 0.05% so that excess water, which may be harmful to crops, can drain easily. Even basins are sometimes graded gently to assist drainage.

(a)

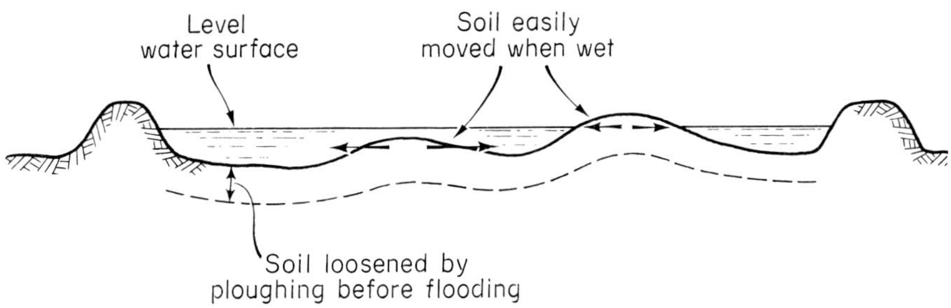

7.6 Levelling rice basins. (a) Water levelling (b) On a small farm (c) On a large farm

7.6(b)

7.6(c)

Farming practices
These usually determine the method of irrigation and so influence land preparation as just described. Small farms using hand and animal power tend to use small basins and furrows which require very little preparation. The larger mechanised farms often use long furrows and borders which require much more land preparation.

7.3 Levelling rice basins
Land preparation for flooded rice can be quite different to other crops. The basins are often levelled when they are flooded with water. This is sometimes called *water levelling* because the water produces a precise level surface. Any soil rising above the water can be removed and used to fill lower lying areas. If the water is then slowly drained other high spots in the basin will be seen (figure 7.6).

The soil is usually loosened by ploughing before flooding. Once it is wet it flows like mud and so can be moved around the field easily by pushing. On small farms this is done using boards pulled by bullocks. On larger fields tractor drawn implements are used. In many rice growing areas the soil is hard and compact under the mud layer and so it can support the weight of the tractor.

7.4 Equipment
Land preparation for irrigation usually involves the use of special equipment at each of the following stages:

Bush clearance

Grading

Planing

7.7 Bush clearance using crawler tractor.

Bush clearance

When new lands are opened for irrigation complete removal of any bush, trees and roots will be necessary. Small areas of land can be cleared with axes, saws and burning. Large areas are usually cleared by machines. The most common machine used for trees and dense bush is the bulldozer. This is a large crawler tractor fitted with a blade for pushing over trees (figure 7.7). Other blade attachments are available for digging out tree stumps and roots. The debris is usually pushed together and burnt.

Light bush can be cleared using a ball and chain. A very heavy steel chain with a large steel ball in the middle is fastened between two crawler tractors and pulled through the bush.

Bush clearance is usually followed by deep loosening of the soil to remove any tree roots and to break up compacted soil layers. Large crawler tractors are needed for this work to pull large steel blades (sometimes called sub-soilers) mounted on tool bars through the soil.

Soil can be loosened down to 700 mm with this method.

Grading

Bulldozers are also used for moving soil over short distances (less than 100 m). They push the soil with a large steel blade. Although this is a very cheap method of grading it is not very accurate (figure 7.8).

For longer hauls it is usually better to pick up the soil and carry it using a motor scraper. This is a wheeled or crawler tractor pulling a large bucket or scraper which can excavate and carry soil (figure 7.8). Wheeled scrapers are faster than crawlers but they have less traction. A second crawler tractor is often needed to push the scraper when excavating. They can carry as much as 10-30 m^3 of soil at a

7.8 Transporting soil. (a) Bulldozer (b) Motor scraper.

7.8(b)

time. Special hydraulic systems are fitted so that the operator can control the depth of cut during excavation and the depth of fill during unloading. These machines are only suitable for use on areas larger than 4 ha. They can prepare land more accurately than bulldozers but are still not accurate enough for surface irrigation. The land may vary up to ± 100 mm from the desired surface (figure 7.9). Clearly if the land is already within this tolerance the use of motor scrapers to try and improve grades may only make the matter worse rather than better.

Planing

More accurate land preparation is achieved with motor graders and land planes (figure 7.10). Although much depends on the skill of the machine operator land can be prepared to within ± 25 mm of the desired grade (figure 7.9). This is suitable for most surface irrigation work.

These machines are used for the final smoothing work. They consist of a long wheeled frame about 8 m long. A blade is located about two thirds of the way along the frame which is tilted forward at the top, not backwards like a bulldozer. The front is hitched to a tractor where the depth of cut can be set. The rear wheels run on the planed surface. The soil is pulverised as it rolls along in front of the blade to produce a good tilth for seed drilling or planting.

A typical method of land planing is to pass over each part of a field three times (figure 7.10). The first and second passes are diagonally across the field and the third is in the direction of irrigation.

7.9(a)

7.9(b)

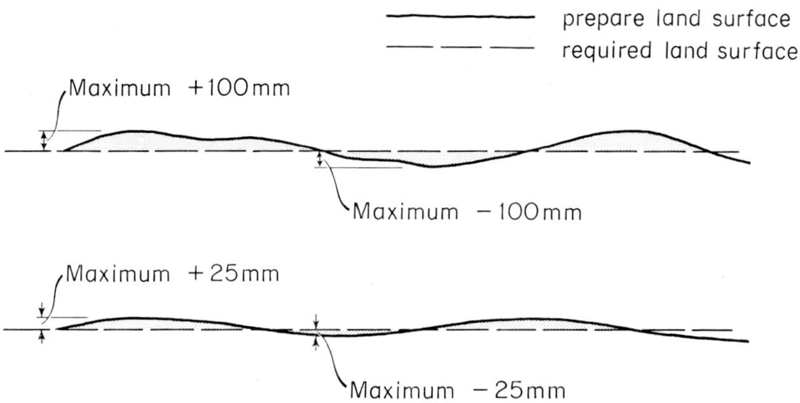

7.9 *Tolerances on land preparation. (a) Motor scrapers (b) Motor graders and land planes*

7.10(a)

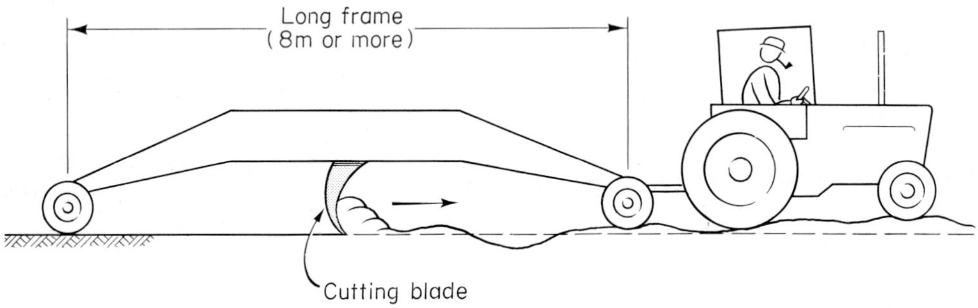

7.10 *Land planing. (a) Land plane details (b) Motor grader (c) Land plane (d) Typical work pattern*

7.10(b)

7.10(c)

Work rates

The rate of working depends on the type of equipment and the land to be prepared. However, a typical job involving one motor scraper and two land planes could prepare up to 4 ha in one day. This assumes about 250-300 m^3 of earth moved per hectare of land.

Using lasers

This is a recent development in land preparation allowing machines to work faster and more accurately.

A laser produces a fine parallel beam of light which can be used to control grading equipment automatically. It is mounted on a tripod in the field (figure 7.11). The laser slowly rotates and the beam of light is set so that its plane of rotation is parallel to the gradient required on the land. Towers attached to the cutting blade on the machine

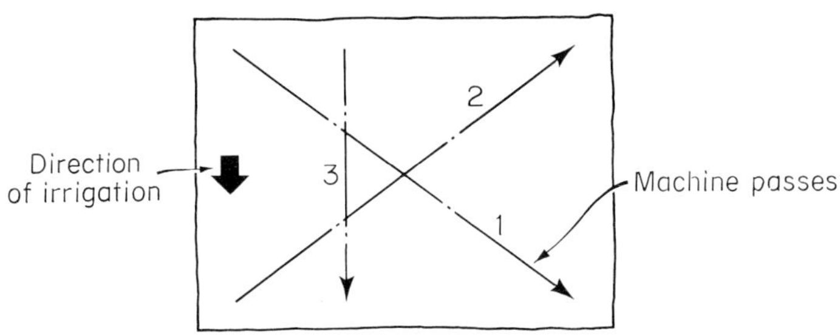

7.10(d)

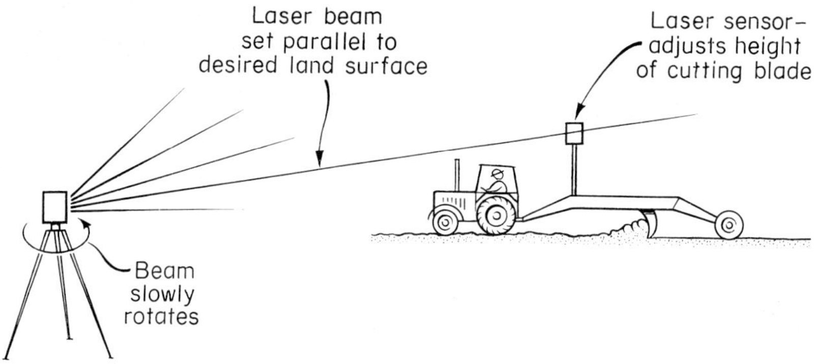

7.11(a)

intercept the laser beam and automatically adjust the position of the blade so that it is at the correct level. This means that as the machine moves over the humps and hollows in the field the blade automatically cuts soil above the desired grade and fills areas below it. This equipment can be used on bulldozers, motor scrapers, graders and land planes.

Land can be prepared up to four times faster with this method. There is also no need to re-survey the land after grading because the operator can be sure that the work has been accurately done when no more soil is being moved.

7.11 *Using lasers for land preparation. (a) General layout of laser system (b) Laser controlled land grading*

7.11(b)

7.5 Common problems

Soil settlement

When irrigating newly prepared land for the first time settlement can occur in the hollows which have been filled with soil. This results in minor irregularities in the soil surface even on land which has been carefully graded. An example of this is shown in figure 7.12. Although the land was carefully levelled with a laser-controlled land plane, the soil has settled considerably near the farm channel.

Clearly land preparation is not something which is done once only, like building a structure. It must be repeated each season until a good land surface is obtained. For this reason it is best to plant only annual crops in the first few seasons. Perennial crops can be planted when all the settlement is complete.

Wet soil conditions

Unlike rice basins land should only be prepared when it is relatively dry. Trying to move soil when it is wet and muddy will not only be difficult but it can cause soil compaction and structural breakdown of the soil. This will inhibit root growth and so reduce crop yields. A good time to prepare land is during the dry season or shortly after a crop has been harvested.

7.12 Common problem – settlement in basins even after levelling using lasers.

8

OPEN CHANNELS

8.1 Introduction

Open channels are the most common method of conveying water. In the irrigation system they are usually called *canals*. In the drainage system they are referred to as *drains*.

Water will only flow downhill in open channels. For this reason canals are usually located along high ground so that water will flow down through the canals and into the fields (figure 8.1). Drains are located along low ground and collect together excess water from the fields for disposal. When the land slopes uniformly, canals and drains can be straight and parallel to each other. On

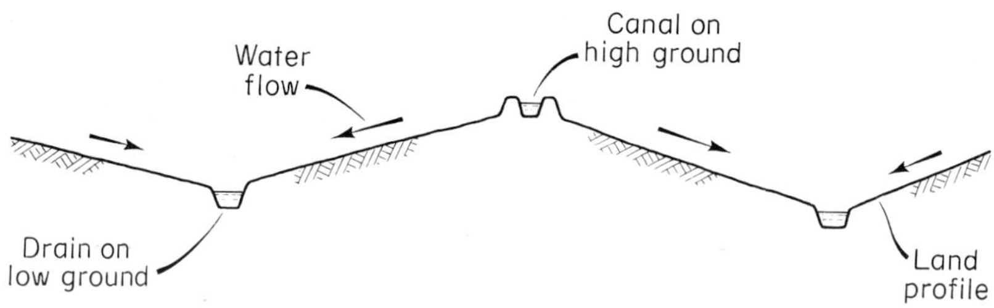

8.1 Locating canals and drains.

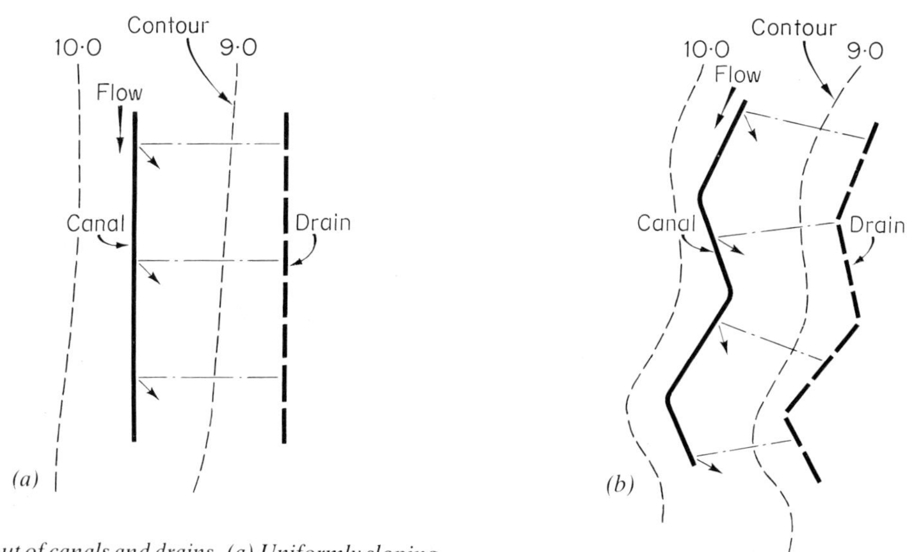

8.2 Layout of canals and drains. (a) Uniformly sloping land (b) Undulating land

undulating land the channels must follow the shape of the land for the flow to continue downhill (figure 8.2).

On a small irrigation scheme there may only be one small canal supplying water to a farm or group of farms and a small drain to dispose of any excess water. On a larger scheme a more complex system of canals and drains are needed (figure 8.3). A typical

8.3 Channels and their location on an irrigation scheme.

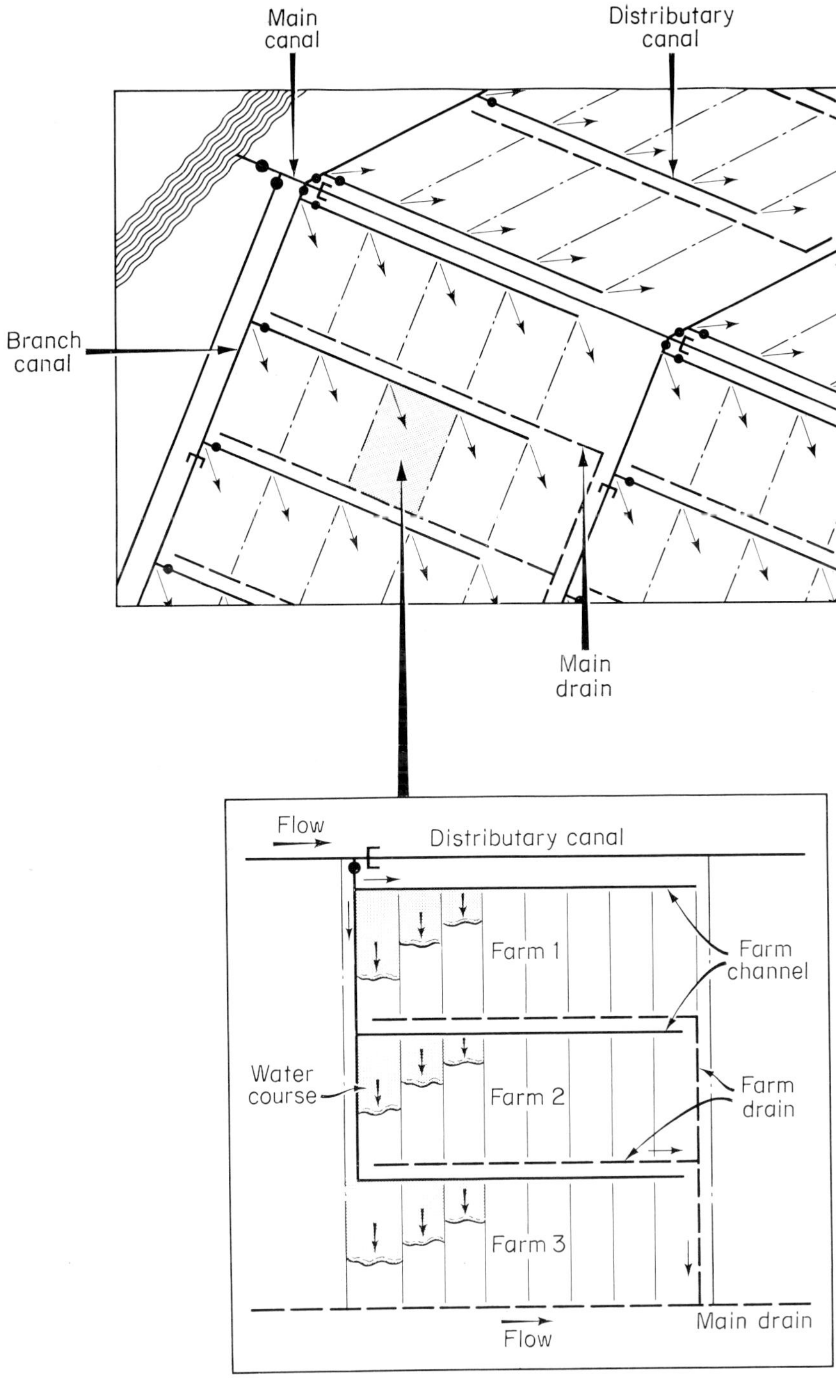

system consists of:

Channel	Function
Main canal	Brings water from source to irrigated area
Branch canal	Delivers water from main canal to parts of the area for distribution
Distributary canal	Distributes water from the branch canal to irrigation units
Water course	Distributes water to farms within irrigation unit
Farm channel	Distributes water within the farm
Farm drain	Drains excess water from the farm
Main drain	Removes excess water from irrigated area

The names given to canals and drains differ greatly from one country to another. In some countries canals are called majors, branches and minors or primary secondary and tertiary canals. The names are different but their function is the same.

All open channels must be large enough to carry the flow of water safely without causing seepage or erosion problems. They must also be cheap to construct and easy to maintain. The design of channels can be a complex task, and is usually carried out by an experienced civil engineer.

CANALS

8.2 Main, branch and distributary canals

These are permanent canals used to distribute water in any large irrigation scheme.

Typical features

Canals have many features in common (figure 8.4). The waterway is usually trapezoidal in shape although some channels are rectangular or semi-circular. It must be large enough to carry the discharge needed. The bed width, water depth and velocity are usually chosen by an experienced engineer.

To stop water spilling over and washing away the canal banks they are usually built much higher than the normal water level. This is called the *freeboard.* On a small water course the freeboard may be only 100 mm but on larger canals a freeboard of 0.3 m – 0.7 m is needed. This is because a breech on a small channel may not be a serious problem and can easily be repaired. A breech on a large canal can cause a lot of damage to farms and may take many days to repair and cost a great deal.

Canal bank tops are usually graded so that rain water does not collect and cause erosion. They are usually wide enough for vehicles. When there is a lot of traffic the banks need to be at least 5 m wide. For maintenance traffic they need only be 4.0 m wide. On some schemes heavy lorries carrying crops from the farms have caused damage to canal banks and in particular to canal structures. To stop this the bank tops can be made too narrow for the lorries. The best access for heavy vehicles of

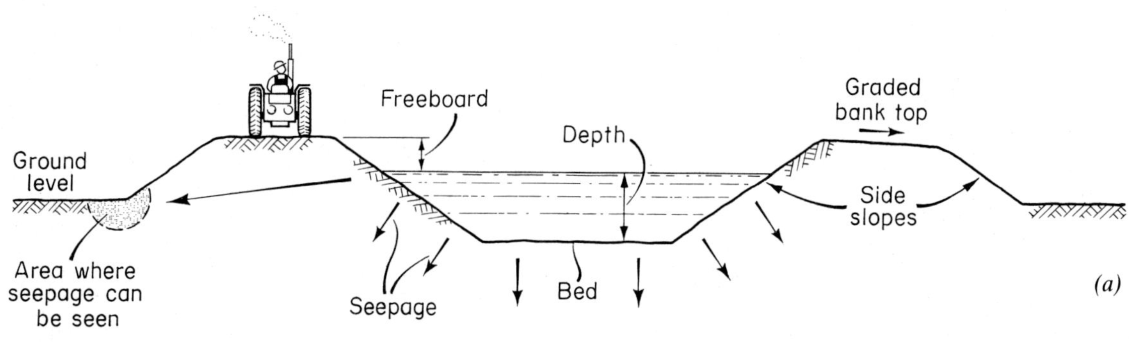

8.4 Typical canal features. (a) Unlined canal (b) Lined canal

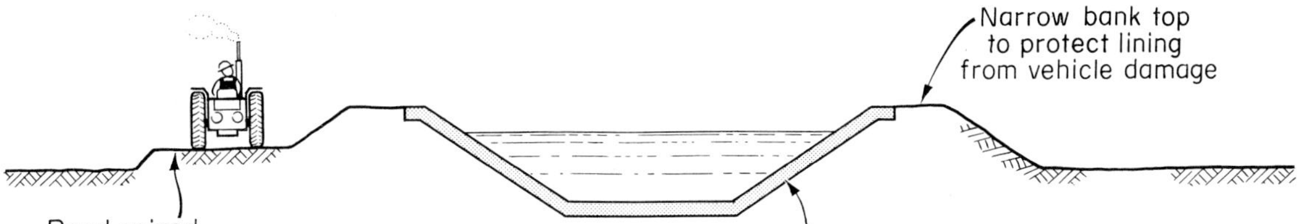

8.4(b)

this kind and farm machinery is along the drainage system where they can do no harm. Narrow bank tops are also used to protect canal linings from vehicle damage. In such cases access roads are built alongside the canal (figure 8.4). These need to be higher than the surrounding irrigated land so that they remain dry.

Water seepage through the bed and sides of a canal is a problem with unlined canals. Seepage can often be seen at the base of canal banks (figure 8.4). A wet muddy area develops in which plants will not grow and vehicles cannot pass. It is a good breeding place for snails and mosquitos and so it may also create a disease hazard (chapter 13). Canal banks must be wide enough to stop this leakage from occurring.

Side slopes both inside and outside the canal are chosen so that the banks always remain stable. If they were too steep they would be continually sliding and this would eventually destroy the canal bank. Typical side slopes vary from 1 in 1.5 for clays to 1 in 3 for sandy soils.

Unlined canals

This is the most common type of canal (figure 8.5). The channel is excavated in the natural soil and the material is used to form the embankments. This is the cheapest and easiest type of canal to construct and is best suited to flat or gently sloping land with very low infiltration rates.

Water velocities are generally low to avoid eroding the banks and sides of the canals. In

8.5 Unlined main canal.

sandy soils the velocity is limited to 0.3 m/s and on clay soils it is 0.75 m/s. When land slopes are steep it can be difficult to keep velocities below these values. In such cases, fall structures are used to break the slope into several gently sloping steps to stop erosion. (chapter 9).

Some unlined canals are specially designed to carry clear water from storage reservoirs and dams. Others are designed to carry silty water from rivers. This means that canals need to flow slow enough to avoid any erosion of the bed and sides, but not so slow that any silt in the water can settle out and block them. Ideally silt is transported along the canals and onto the fields where it becomes a good source of nutrients for crops. This does not always happen in practice and so regular cleaning is needed. Silt also encourages weeds and reeds to grow in the canals and these need to be cleared regularly for water to flow properly. Aquatic weeds can also provide good breeding places for snails and mosquitos which can spread disease (chapter 13). So, although unlined canals are cheap and easy to construct they do need a lot of maintenance.

Lined canals

Canals are lined for two main reasons:

To reduce seepage losses

To reduce maintenance

Various materials are available to line the waterway to reduce seepage. The cheapest is a layer of clay soil placed and compacted on the bed and sides of the canal (figure 8.6). This will reduce seepage but does not stop weed growth and soil erosion problems.

Hard-surface linings such as concrete, brick and bitumen reduce seepage and also stop weed growth and erosion (figure 8.7). This greatly reduces the maintenance required. Concrete lining is the most durable of all the lining materials. It can be laid as pre-cast slabs or cast in place. Special machines have been built to excavate canals and line them with concrete (figure 8.7). These are useful when there is a lot of lining to be done.

Membrane linings such as pvc or butyl rubber are also used to reduce seepage. They are usually buried below the bed of the canal to protect them from damage by spades and other sharp implements. For this reason they do not reduce maintenance work because the weeds can still grow.

Some countries choose to line canals specially to reduce maintenance. This is because of the problems they face when trying to maintain unlined canals. (chapter 12). A hard-surface lining will reduce weed growth and erosion damage initially but it will not last for ever. It will slowly deteriorate and need replacing and this can be expensive. The life of a lining depends on the material used and on how well it was installed. For example, a concrete lining may last 50 years or more when properly installed. But if the embankments underneath the lining have not been properly compacted or the concrete poorly vibrated the lining may crack within a year or two. This will result in a great deal of maintenance work of a very skilled type and will be very expensive to repair.

Lined canals should only be used to carry silt-free water because it is difficult to remove sediment without damaging the lining. If irrigation water is silty then it is usually passed

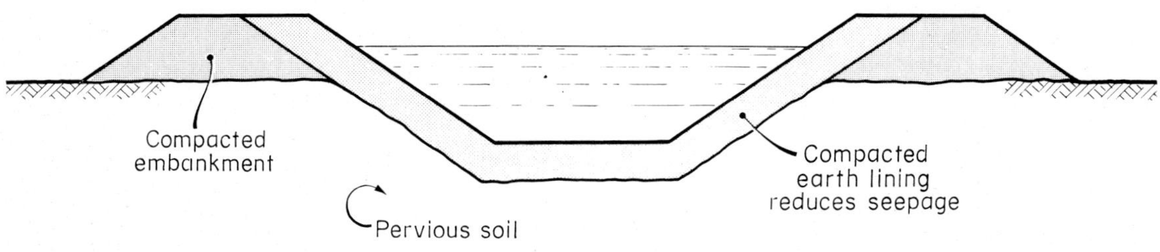

8.6 Clay-lined canal.

(a)

(b)

8.7 *Concrete-lined canals. (a) Main canal (b) Lining machines*

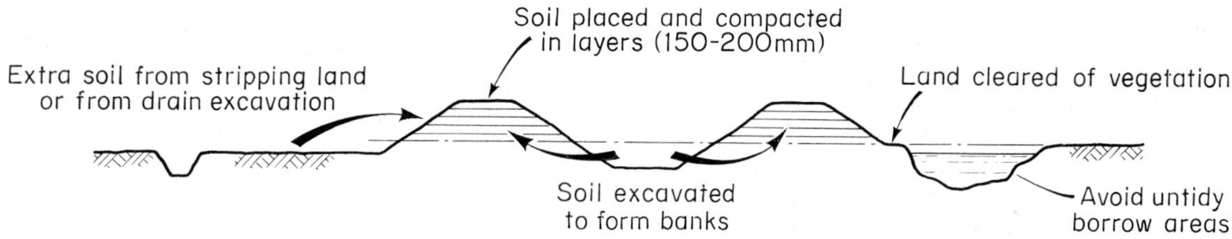

8.8 Canal construction.

throught a settling basin to remove the silt before the water flows into the canal system.

Construction

First the position of the canal is set out on the ground and the land cleared of trees, shrubs, grass and vegetation.

Soil is then excavated from the canal water way and placed on the embankments (figure 8.8). Any extra soil needed is taken from areas close to the canal. This is best done by stripping a thin layer of soil from a wide area or using soil excavated from drain channels. On some schemes soil is excavated alongside the canals and large holes are formed as a result. These are called *borrow-pits* and they can cause a lot of problems. Not only can they look untidy, they fill with water and if not drained properly they can be a health hazard. (chapter 13)

Canal embankments are constructed in layers of 150-200 mm, building up to the full height. This is done so that the soil can be properly compacted to reduce water seepage. Some settlement of canal banks can be expected and so they are usually built higher than is needed to allow for this. Good compaction of soil in canal banks is very important when canals are to be lined. This will avoid settlement and damage to the lining in the future.

Canals are usually constructed using heavy earth moving machinery. Bulldozers are sometimes used, but they do not compact the material properly. Large canals are usually constructed with motor scrapers to transport the soil and graders to spread it. This equipment is similar to that used in land preparation (chapter 7).

8.3 Water courses and farm channels

These channels are much smaller than the canals in the irrigation system. They are used to distribute water within the irrigation unit. Like canals they can be lined or unlined depending on seepage, erosion and maintenance problems. Generally they are triangular, trapezoidal or rectangular in shape (figure 8.9).

Farm channels only need small banks to contain the flow because there is not much risk involved if the canal bank breaks. Indeed on many schemes, irrigators open the banks to release water into the fields, although this is considered bad practice.

(a)

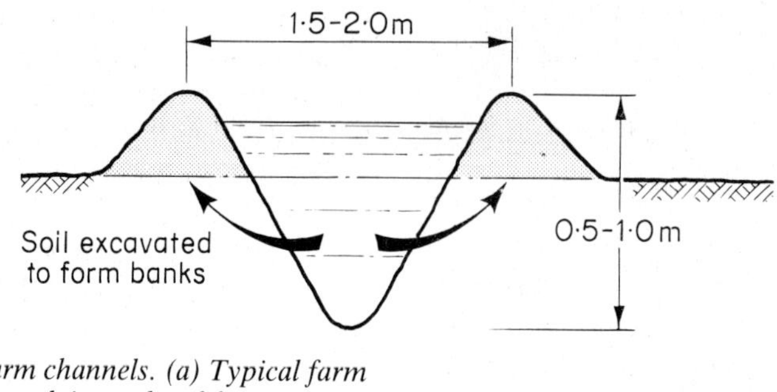

8.9 Water courses and farm channels. (a) Typical farm channel (b) Watercourse supplying unlined farm channels (c) Lined rectangular watercourse

8.9(b)

8.9(c)

Construction

Channel banks are usually formed from soil excavated from the water way. In many cases the size of the channel is determined by the amount of soil needed to build the banks rather than being able to carry the farm discharge. There are special ploughs which can form the complete channel as it is pulled along by a tractor. In places where farms are small and labour is plentiful the channels can be dug by hand (figure 8.10). The small banks will need compacting to seal them and stop seepage.

8.4 Command

If water is to flow onto the fields without pumping the water level in the canals must be above ground level. The difference in level between the water surface and the ground is called the *command* (figure 8.11).

The command in a farm channel should be sufficient for water to flow easily onto the field. For example, if the depth of water ponded in a basin is 75 mm and the head required to operate the siphons is 100 mm the command in the farm channel should be 175 mm. If it is below this, the water level can be raised by building a check structure across the channel (figure 9.10).

Water levels in distributary, branch and main canals can be determined in the same way. In the distributary canal the water level must be high enough to command each irrigation unit. Typical commands are 300-500 mm. This allows for losses in head as water flows through the canal outlet, measuring and flow diversion structures (chapter 9).

When the command is low the flow from the canals to the fields is reduced. When it is below ground level pumping will be needed. In Egypt most canals operate with no command. Irrigators must pump the water and so they are careful not to waste any.

When the command is too high, there may be seepage through the canal banks and erosion in the fields when irrigating. The cost of constructing the canals also increases.

8.10(a)

8.5 Seepage

This is a common source of water loss in unlined canals and in some lined canals. Tests can be carried out on canals to measure just how much water is being lost in this way. This will help irrigators to locate and reduce losses and help designers to decide whether to use lining on any new canals.

Two of the most common methods are:

Ponding method

Inflow-outflow method

Ponding method

This method involves blocking off a length of canal and filling it with water. The drop in water level over a period of time is the amount of water lost from seepage (figure 8.12).

The test is best done when the canal is in use and the results are typical of normal operating conditions. A gauge is placed at each end of the canal and the fall in water level over a 24 hour period is measured. The average of the two readings is taken to eliminate any wind effects on the water surface. Seepage along each kilometre of a small channel can be calculated approximately by:

$$\underset{(\ell/s/km)}{\text{Seepage}} = 0.013 \times \text{water surface width (m)} \times \text{drop in water level over 24 h (mm)}$$

This test can be difficult to do on a large canal because it involves closing the canal and

8.10(b)

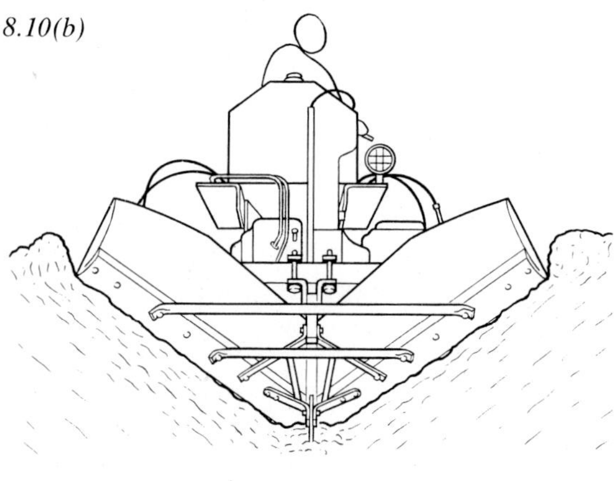

8.10 Watercourse and farm channel construction. (a) Excavating channel by hand (b) Farm channel formed by 'vee' plough

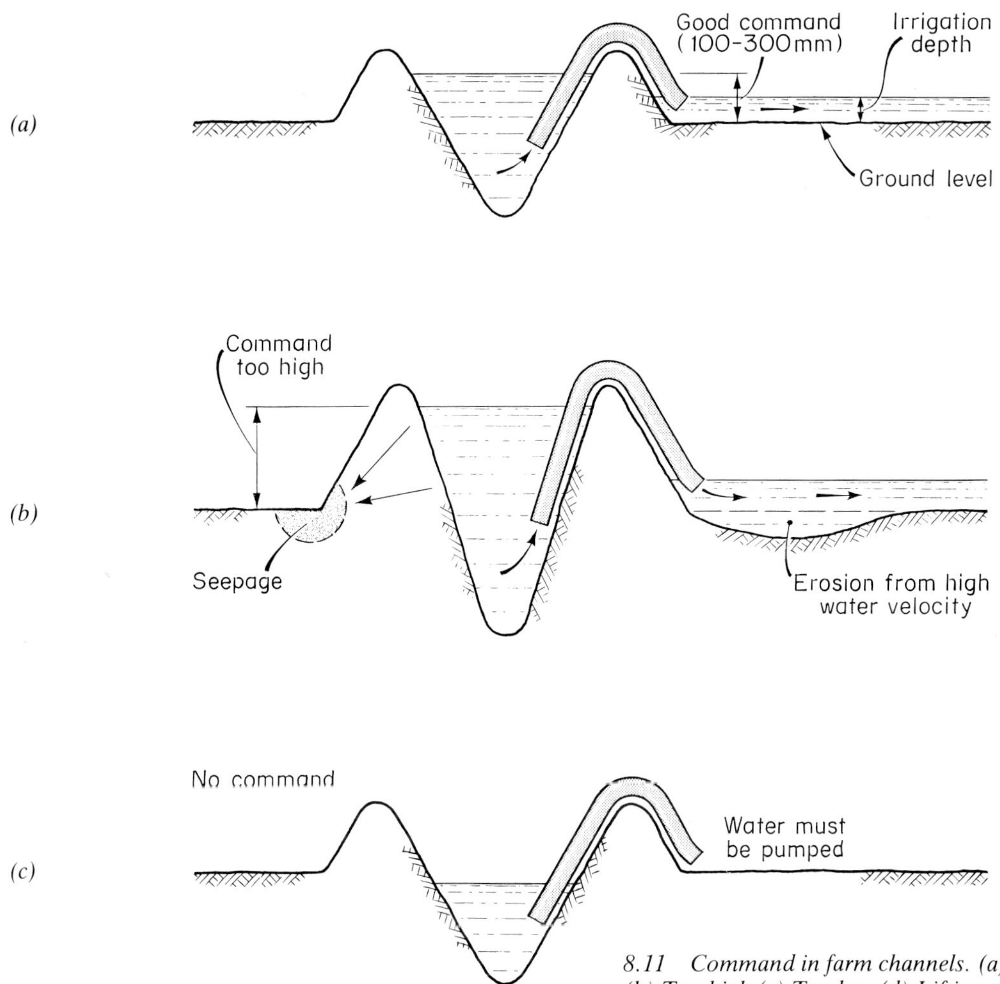

8.11 Command in farm channels. (a) Right command (b) Too high (c) Too low (d) Lifting water when command is too low (Egypt)

measuring any leaks from structures. Small farm channels can be tested easily without too much interruption to irrigation. It is also easy to locate and block any leakage which would upset the results.

Inflow-outflow method

This method involves measuring the flow in a canal at two places along its length. The difference between the two flows is the amount of water lost from seepage.

The seepage for each kilometre of canal can be calculated as follows:

$$\text{Seepage } (\ell/\text{s/km}) = \frac{\text{inflow } (\ell/\text{s}) - \text{outflow } (\ell/\text{s})}{\text{length of test canal (km)}}$$

Total seepage in the canal can be calculated as follows:

$$\text{Total seepage} = \text{seepage } (\ell/\text{s/km}) \times \text{canal length (km)}$$

When deciding on the discharge required in the canal for irrigation the amount of seepage would need to be added to the flow. On main, branch and distributary canals, hydraulic structures can be used to measure the discharge, (see chapter 9). If they cannot be used to measure the flow accurately enough then current metering or the float method can be used. (section 2.4)

On watercourses and farm channels rectangular weirs temporarily installed can be

8.11(d)

8.12(a)

8.12 *Measuring seepage in canals. (a) Ponding method (b) Inflow-outflow method*

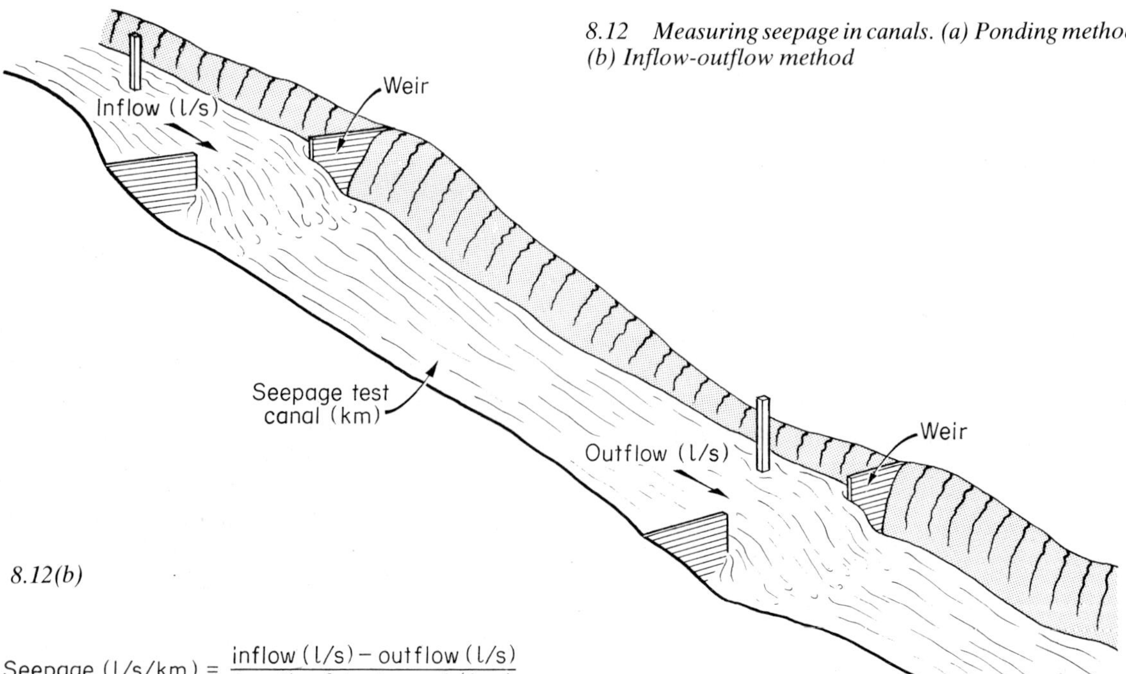

8.12(b)

$$\text{Seepage (l/s/km)} = \frac{\text{inflow (l/s)} - \text{outflow (l/s)}}{\text{length of test canal (km)}}$$

used to measure inflow and outflow. It is best to choose a length of canal from which there is no abstraction taking place.

DRAINS

Farm drains

These channels remove excess water from farms as a result of too much rainfall or irrigation.

Surface water is usually collected by a system of shallow drainage channels along the lower ends of the farms (figure 8.13). Gentle side slopes are used so that farm vehicles can drive across them to gain access to the fields.

On some schemes much deeper drains are needed to drain the ground-water as well as surface water (figure 8.13). Soil excavated from the drain is used to form embankments to protect the channel from erosion by surface water flow. Berms are left so that excavators can get easy access for maintenance.

Main drains

These are very similar to the farm drains in appearance but are usually larger because they must carry water from many farms. The main drains remove water from the irrigated area for disposal.

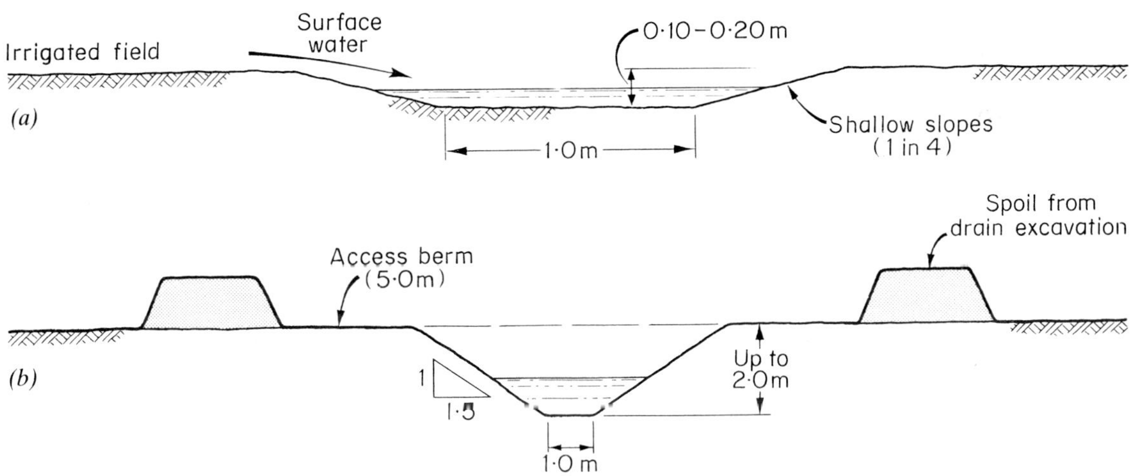

8.13 Farm drains. (a) Surface-water drain (b) Ground water drain (c) Drain excavation using draglines

(c)

9

HYDRAULIC STRUCTURES

9.1 Introduction

Hydraulic structures are used in canals for three reasons:

- To measure and control the distribution of water
- To maintain water levels for command
- To dissipate unwanted energy.

There are many kinds of hydraulic structures. Some of the more common ones used on irrigation schemes are described below. Their typical location is shown in figure 9.1.

9.2 Head regulators

These structures are used to control and measure the flow of water into canals. On a small scheme there may only be one head regulator. On a large scheme there would be one at the head of each main, branch and distributary canal.

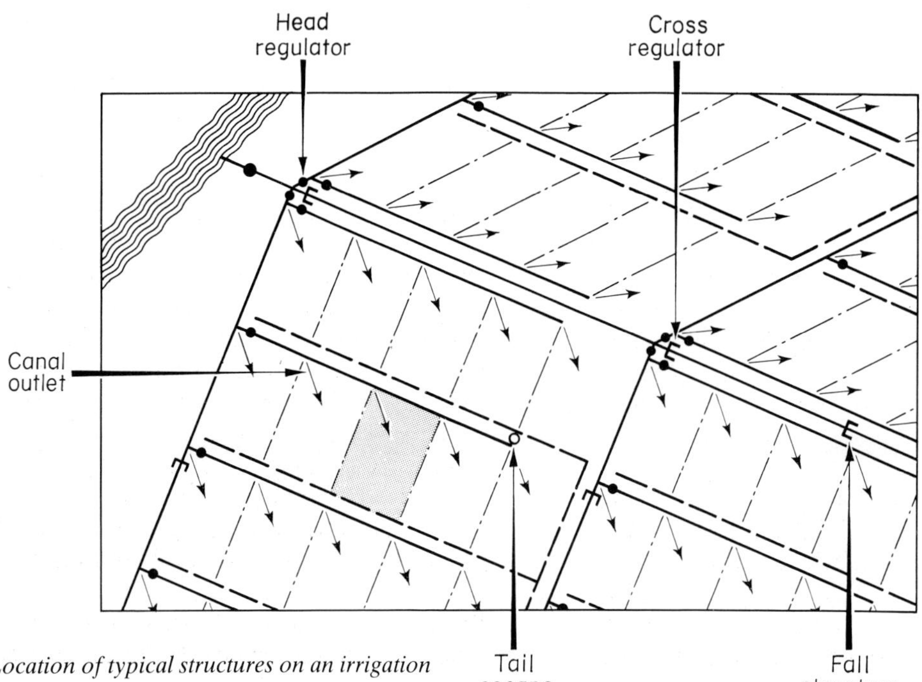

9.1 Location of typical structures on an irrigation scheme

(a)

(b)

9.2 Head regulators. (a) Lifting gate (b) Movable weir

Large canals carrying high discharges require large regulating structures. These are usually built from concrete and the flow is controlled by vertical lifting gates or radial gates (figure 9.2). The gates are raised or lowered into the water by hand operated winches or electric motors if a power supply is available.

On smaller schemes and on the smaller distributary canals much smaller structures are used. These can be vertical lifting gates or moveable weirs (figure 9.2). Although concrete can be used to build these small structures, they are often constructed from blocks, bricks or stone if it is available locally. Moveable weirs are very useful regulators. The gate is shaped like a weir and the discharge is measured by observing the depth of flow over the weir crest. The gate is raised or lowered to adjust the flow into the canal.

9.3 Cross regulators

These look very similar to head regulators but they are used for a different purpose. When a canal is operated below its normal discharge water levels are usually lower and so the command is reduced. In some cases the command can be so low that water has to be pumped from the canal. To overcome this problem structures are built across the canal to maintain the water level at the command required to irrigate the fields. These structures are called *cross regulators*.

On a large irrigation scheme cross regulators are located close to head regulators to control the water levels at canal junctions. One cross regulator can control the water level for several head regulators. Together they form a regulator group (figure 9.3).

Automatic gates can be used for cross regulation (figure 9.3). They automatically control a constant water level in the canal for different discharges. Although they require very little attention from canal operators they are expensive to buy. Regulation can also be achieved with very simple brick structures and wooden gates. The gates are raised and lowered using block and tackle from an overhead beam.

Cross regulators are also used on distributary canals where water is taken from the canal into the irrigation unit. This is usually a much smaller and simpler structure than is used in the main and branch canals. Small lengths of wood called *stop-logs* can be fitted across the structure to obtain the correct water level in the canal.

9.3 Cross regulators. (a) Typical regulator group
(b) Automatic gate (c) Simple brick and timber regulator
(d) Small regulator on a distributary canal

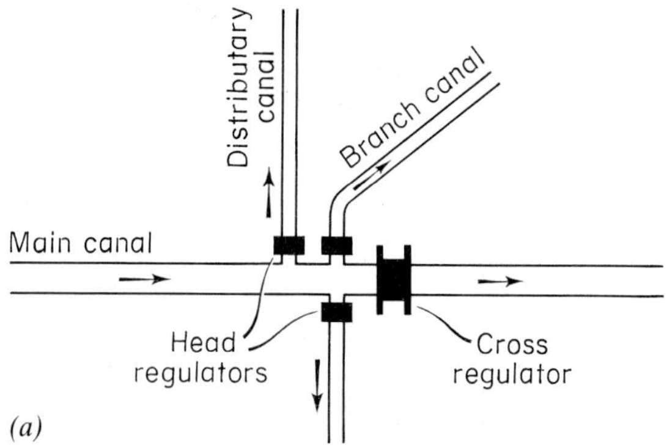

(a)

(b)

(c)

(d)

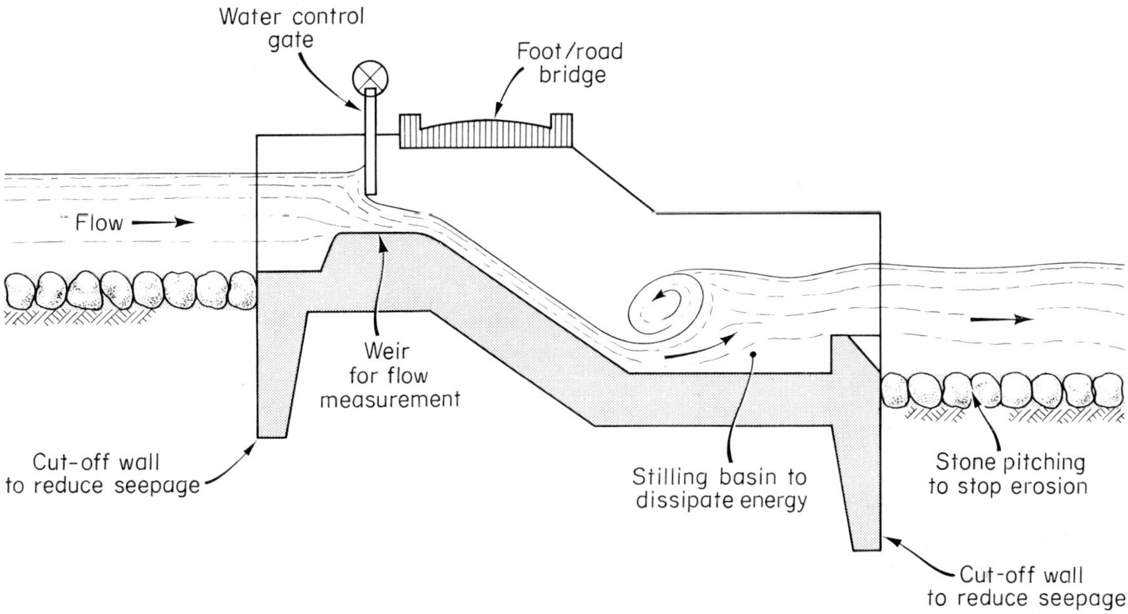

9.4 Components of typical canal regulator.

9.4 Components of a typical regulator
Although there are many different types of regulator they all have the following in common (figure 9.4):

A weir to measure the discharge. Some regulators are fitted with gates to improve control over the flow.

Special walls called *cut-offs* to reduce seepage of water underneath the regulator.

A special area downstream called a stilling-basin in which excess water energy can be safely dissipated.

Concrete or stone blocks upstream and downstream to protect the channel from erosion.

A bridge for general access.

9.5 Fall structures
When unlined canals are constructed on steeply sloping land water velocities can be high and there is the risk of erosion. To overcome this problem the designer usually chooses gentle sloping lengths of canal with fall structures in between (figure 9.5). This is rather like a staircase. Water flows along the canal then falls down to the next level. The fall structure allows this to happen safely and the water energy is dissipated within the structure.

On many schemes falls are combined with head and cross regulators and so one structure carries out both tasks. Figure 9.3[d] shows a cross regulator combined with a fall. Note the turbulent water where the water energy is being dissipated.

9.6 Tail escapes
Most canals need some way of getting rid of excess water. This may be due to too much water flowing into the canal from the head regulator or irrigators closing their canal outlets. A structure can be provided at the end of the canal so that excess water can flow safely into the drains. This is called a tail escape (figure 9.6). It consists of a fall structure combined with a pipe culvert under the canal embankment.

(a)

9.5 *Fall structures. (a) On a main canal (b) On a distributary canal* (b)

(a)

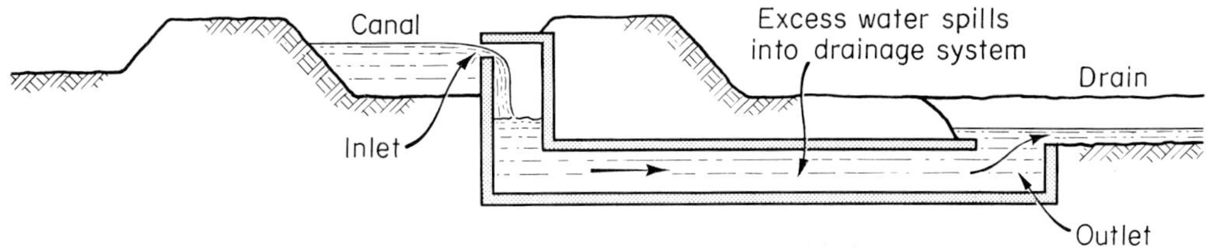

(b)

9.6 Tail escapes. (a) General detail (b) Inlet at end of distributary canal

9.7 Canal outlets

This is usually the last control structure in the irrigation system and is perhaps the most important. It controls the flow of water to the irrigation unit. It is in fact a kind of head regulator but it is generally called a *canal outlet.*

There are many different types of canal outlet in use in many different countries. However, they all have the following in common:

They are used to control and sometimes measure flow into irrigation units.

Each scheme usually has the same type of outlet. This is done to reduce construction costs and to make them easier to use.

They should be tamper-proof. This is to stop some irrigators trying to get more than their fair share of the water.

The simplest canal outlet consist of an opening or pipe through the distributary canal bank (figure 9.7). A lifting gate on the upstream side of the outlet controls flow into the irrigation unit. Some gates simply open or close the outlet, others are fitted with screw shafts and handwheels so that the flow can be adjusted.

Another outlet has a lift gate on the upstream side and a weir to measure the flow on the downstream side (figure 9.7[c]). This provides much better control over the distribution of water but the structure is usually more costly.

Both the canal outlets described are affected by changes in the canal water levels. When the water levels rise and fall the discharge also rises and falls.

A third type of outlet overcomes this problem. It is fitted with a special water control gate which provides a constant flow even though the water levels in the canal may change (figure 9.7[d]). The flow can be varied by opening the different sized gates. Notice the pad locks used to prevent unauthorised use of the outlet.

9.7 Canal outlets. (a) Opening in canal bank

(b) Pipe outlet with adjustable gate

(c) Pipe and weir outlet — upstream side

— downstream side

9.7(d) *Constant flow outlet*

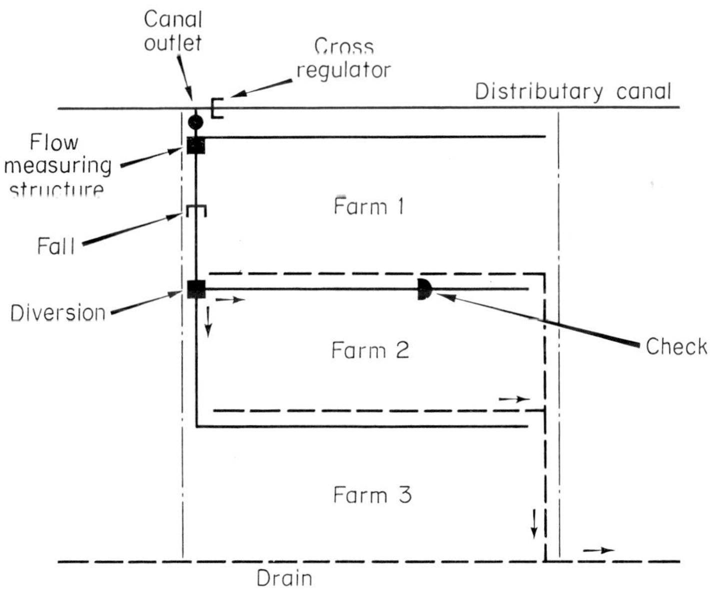

9.8 *Location of typical structures in an irrigation unit.*

9.8 Unit structures

Within irrigation units structures are used to measure the flow in the watercourses and farm channels; to divert the flow from one channel to another and to release water from the farm channels into the fields. Typical locations for the structures described below are shown in figure 9.8.

Flow measuring structures

On some schemes flow measurement is combined with the canal outlet. On other schemes weirs or flumes are installed in the watercourses and farm channels. These are normally conveniently positioned so that they can also be used for other purposes such as diverting flows and cross regulation.

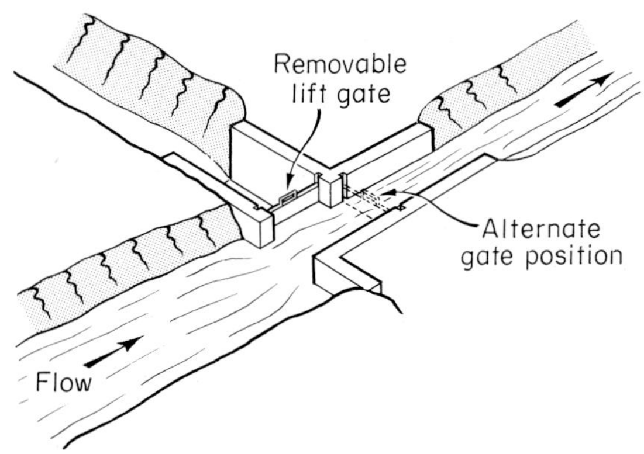

9.9 Flow diversion structure.

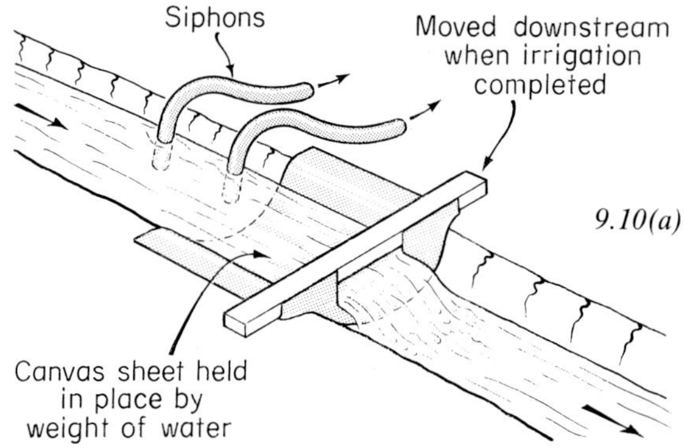

9.10(a)

9.10(b)

9.10 Check structures on farm channels. (a) Portable canvas check (b) Permanent check

Diversion structures

These are small structures used to divert flow from one farm channel to another or to divide the flow between two channels (figure 9.9). They are usually small box shaped structures made from concrete or wood with two or more outlets. Simple lift gates are used to open and close the flow. Sometimes only one gate is provided so that both channels cannot be closed at one time. This can be secured to the concrete with a steel chain if security is a problem.

Checks

Most farm channels need cross-regulators to raise water levels above ground level for irrigation. These structures are often called *checks*. On some channels, permanent structures are installed while on others portable canvas checks can be used (figure 9.10). These are placed in the channel and the weight of water holds the canvas sheet in place. They confine the water to a part of the farm channel being used for irrigation at that time. When the irrigation is completed the check is lifted and moved further down the channel.

Water can be released from the farm channel into the fields through siphon pipes or small structures built into each basin or border (figure 9.11). A common but bad practice is to cut an opening in the bank of a farm channel with a spade. This can destroy the channel banks, leaves a very untidy mess and it becomes difficult to control the flow properly.

Falls

On steeply sloping land falls may be needed in the channels to avoid erosion (figure 9.12). On some schemes irrigators may not fully understand their use. On one scheme the irrigators did not like the fall structures. They appeared to be holding back the water and each reach of the water course took some time to fill before overflowing to the next reach. The irrigators excavated new channels so that the water flowed around the structures. As a result the channel eroded very badly near the top and silted up at the bottom after only a few hours operation. Only by seeing these results did the irrigators realise the importance of fall structures.

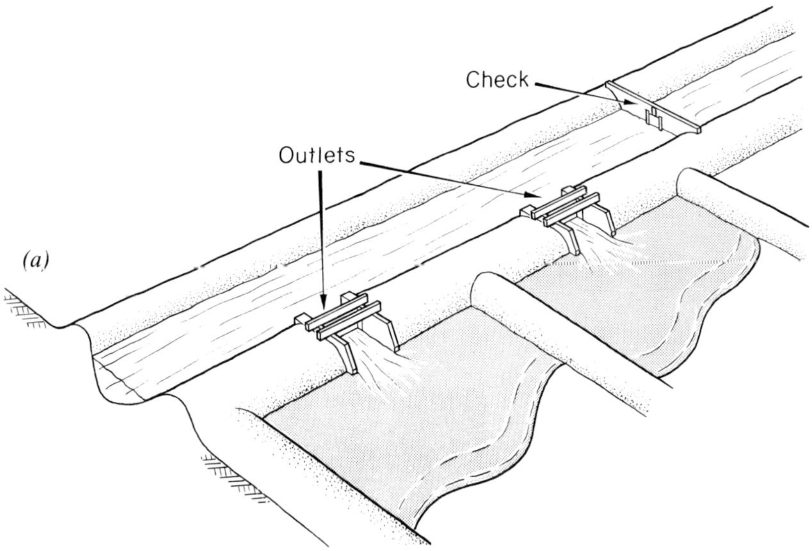

(b)

9.11 Releasing water from a farm channel into the field. (a) Through a small timber structure (b) Cutting a hole in the channel bank

9.12 *Fall structure on a watercourse.*

Whatever structures are used, the channels should always be able to drain down after an irrigation. This will remove any possible breeding places for disease carrying snails and mosquitos.

9.9 River offtake problem

Natural rivers are a very common source of water for irrigation. However, rivers sometimes change their position from year to year. Water levels rise and fall depending on the rainfall. Sometimes the water is very clean and at other times it is very silty. River offtake structures such as head regulators or pumping stations built near rivers are easily destroyed as river banks erode or they become blocked with silt. Also canals can be blocked if too much silt enters the system. Pumps can also be damaged.

Choosing the best site

Very few natural river channels are straight, they tend to meander. When choosing a site for an offtake always look for a position on the outside of a bend in the river (figure 9.13). Rivers always tend to leave silt banks on the inside of bends and scour on the outside.

By siting an offtake on the outside it will usually be free from silting. The outside bend may need to be protected with stone to stop any further erosion which might destroy the structure. If it is possible it is always better to take a channel from the river at a small angle rather than at right angles. This helps to reduce the amount of silt entering the canal when the river is very silty.

Weirs and barrages

When river water levels rise and fall it can be difficult to decide on the best kind of offtake. When the river is low pumping may be needed. When it is high a gravity supply with a head regulator may be suitable.

River levels can be controlled by a special kind of cross regulator. One is a weir constructed across the river close to the

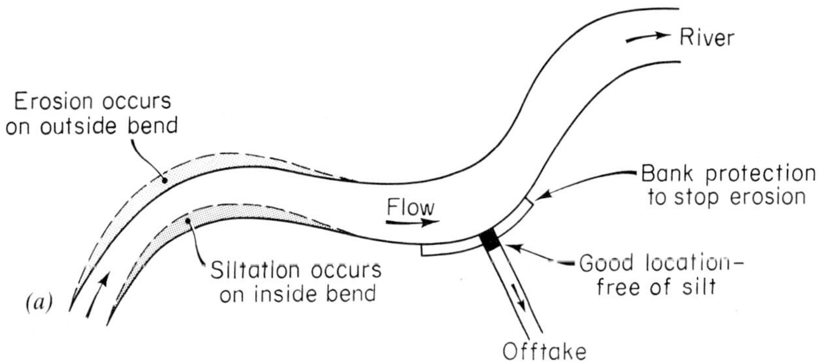

9.13 Choosing the best site for an offtake. (a) Locating offtake on a meandering river (b) Angle of offtake

(a)

9.14 Controlling water levels in a river. (a) Weir (b) Barrage

9.14(b)

offtake (figure 9.14). This keeps the water levels high even when the river flows are very low.

If more control is needed, gates can be built on the weir which can be raised or lowered depending on the flow. This kind of structure is called *a barrage*. The offtakes are usually built near to the barrage (figure 9.14). On some barrages, scour sluices are constructed to stop silting near the offtake. These are lifting gates which are opened occasionally to flush out any silt banks which may have formed.

10

PIPELINES

10.1 Introduction

Pipelines are sometimes used for all or part of an irrigation system instead of open channels. There are many reasons for this:

Reduced seepage and evaporation losses

Less maintenance – no weeds

Better control over water distribution

Can irrigate hilly ground

Less land taken up by the irrigation system

Can improve public health (see chapter 13)

Pipelines have many of the advantages over lined canals, but like lining they can be expensive and skilled labour is needed for their manufacture and installation.

10.2 Types of system

There are three main types of system:

Permanent

Semi-permanent

Portable

Permanent

This is the most common system (figure 10.1). It consists of a mainline and lateral pipes which are usually permanently installed underground. Sometimes a pump is needed to provide the right operating pressure in the system. Valves are used to divert the flow from the mainline into the laterals. Pipe outlets control the flow from the laterals onto the field. One or more outlets can be operated at the same time.

The system can be used for all methods of surface irrigation. When irrigating basins and borders each basin or border usually has its own pipe outlet. Sometimes the pipe outlet supplies water to a farm channel which then supplies several basins or borders using siphons (figure 10.1[b]). Another method is to use a special hydrant on top of each pipe outlet, which distributes water to several furrows (figure 10.1[c]).

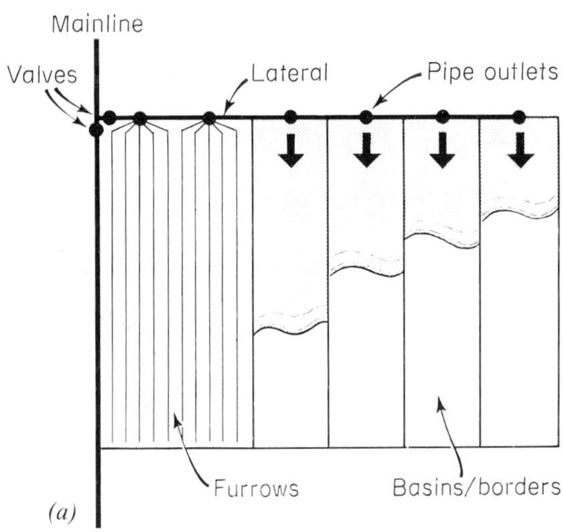

10.1 Permanent system. (a) Typical layout (b) Pipe outlet supplying farm channel (c) Distribution box on pipe outlet for furrows

10.1(b)

10.1(c)

Semi-permanent
This is similar to the permanent system but has fewer pipe outlets (figure 10.2). A portable aluminium or plastic pipe up to 30 m long and 150-300 mm in diameter is connected to one of the pipe outlets and supplies water to several basins, borders or furrows at the same time through small openings along the pipe.

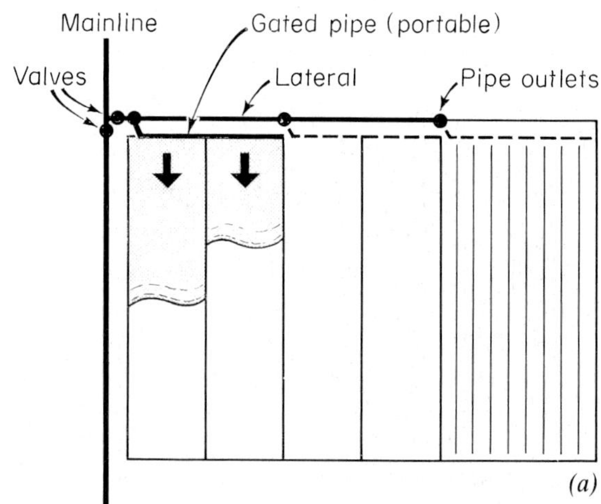

10.2 Semi-permanent system. (a) Typical layout (b) Portable gated-pipe connected to pipe outlet (c) Gated-pipe supplying water to furrows (d) Slide valve fitted in gated-pipe

10.2(b)

10.2(c)

10.2(d)

This is called a *gated-pipe*. Each opening is fitted with a small slide valve. The discharge from each opening can be varied between 0.15-6.0 ℓ/s by adjusting the position of the slide valve. This provides good control over the irrigation stream size. It is very useful in furrow irrigation when using the cut-back method. A large flow is used initially and then cut back by adjusting the slide valve.

When irrigation at the first pipe outlet is completed, the outlet is turned off, the gated-pipes are disconnected and moved to the next outlet. They are light in weight and easily carried. Special couplers developed for sprinkler irrigation systems allow pipes to be connected quickly and simply.

Portable

This is similar to the semi-permanent system but all the pipe work is portable and laid above ground. This means the system can be laid out for each irrigation and moved from field to field. Sometimes it is laid out for the whole season then disconnected and removed into a store for the winter.

SYSTEM COMPONENTS

A pipe system is made up of many components (figure 10.3).

10.3 Mainlines and laterals

These are usually permanent or portable. Permanent pipes are laid above or below ground and are generally made of concrete, plastic, galvanised steel or asbestos cement.

Underground pipes do not use up land space or interfere with farming operations. Normally they are buried with a minimum cover of 0.75 m.

Portable pipes which are moved frequently need to be strong and yet light and easy to handle. They are usually made from aluminium, galvanise steel or plastic.

Pipes used in surface irrigation will safely stand high operating pressures but there can be difficulties if the pressure is too high. Normally a pressure of 2-3 m head of water is sufficient to ensure full flow at the pipe outlet. Higher pressures can cause serious erosion

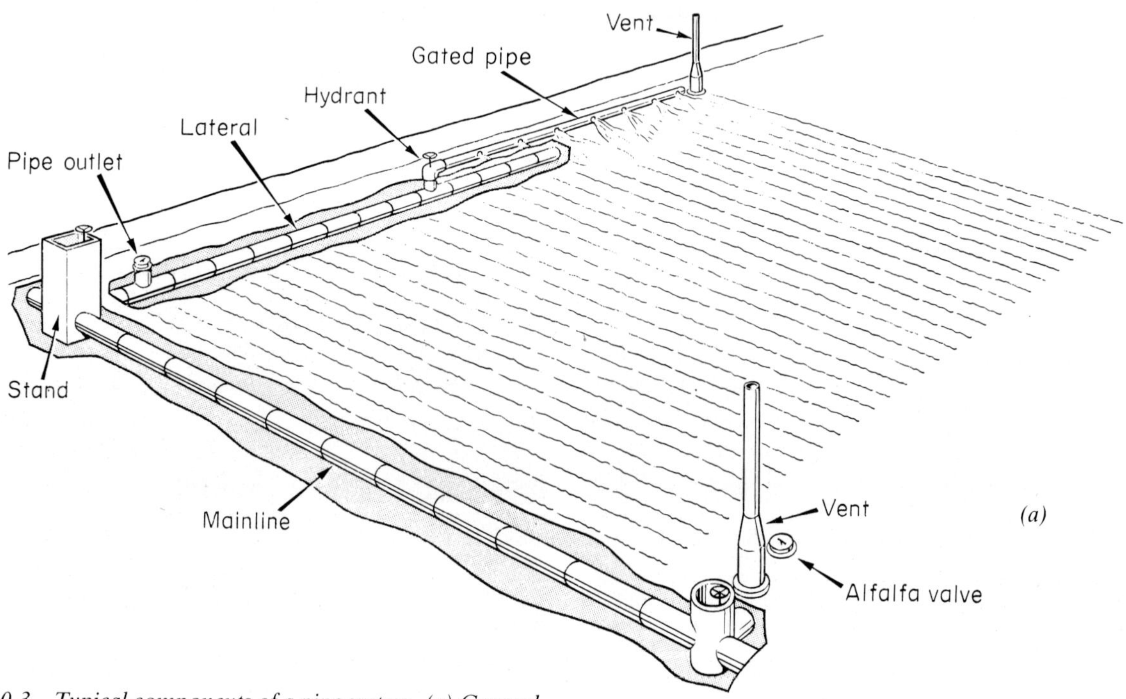

10.3 Typical components of a pipe system. (a) General layout (b) Cover to buried pipes

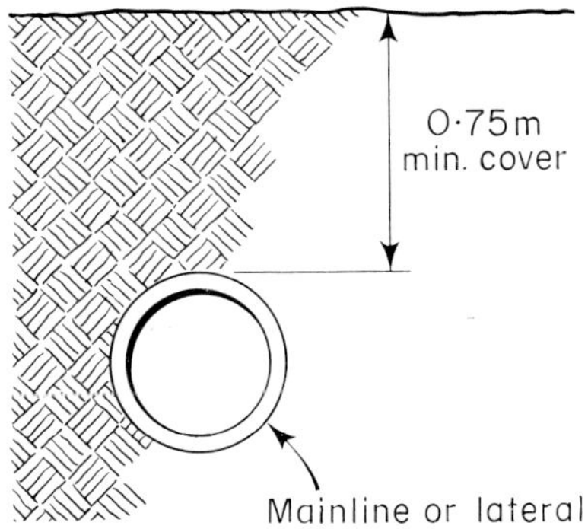

10.3(b)

around the pipe outlets.

Pipes have a maximum discharge capacity at a given operating pressure. The size must be chosen so that it will adequately supply the irrigated area. This is usually done by an experienced engineer.

10.4 Pipe outlets

These are installed at intervals along the lateral to release water into the field. On underground laterals a riser pipe is needed. This brings water to the ground surface. Figure 10.4 shows a typical riser pipe being installed on an unreinforced concrete pipeline. A hole is cut into the top of the lateral pipe. The riser pipe is placed over the hole and the joint sealed with sand-cement mortar. This will be strong enough because the pipes only operate at very low pressures. A valve is placed in the top of the riser to control the water flow. There are two types of valve in common use:

Alfalfa valves

Orchard valves

10.4(a)

10.4(b)

10.4(c)

10.4(d)

10.4(e)

10.4 Installing riser pipe on an underground concrete lateral. (a) Cutting hole in lateral pipe (b) Fitting riser pipe (c) Jointing with sand-cement mortar (d) Fitting valve in top of riser (e) Completed riser pipes

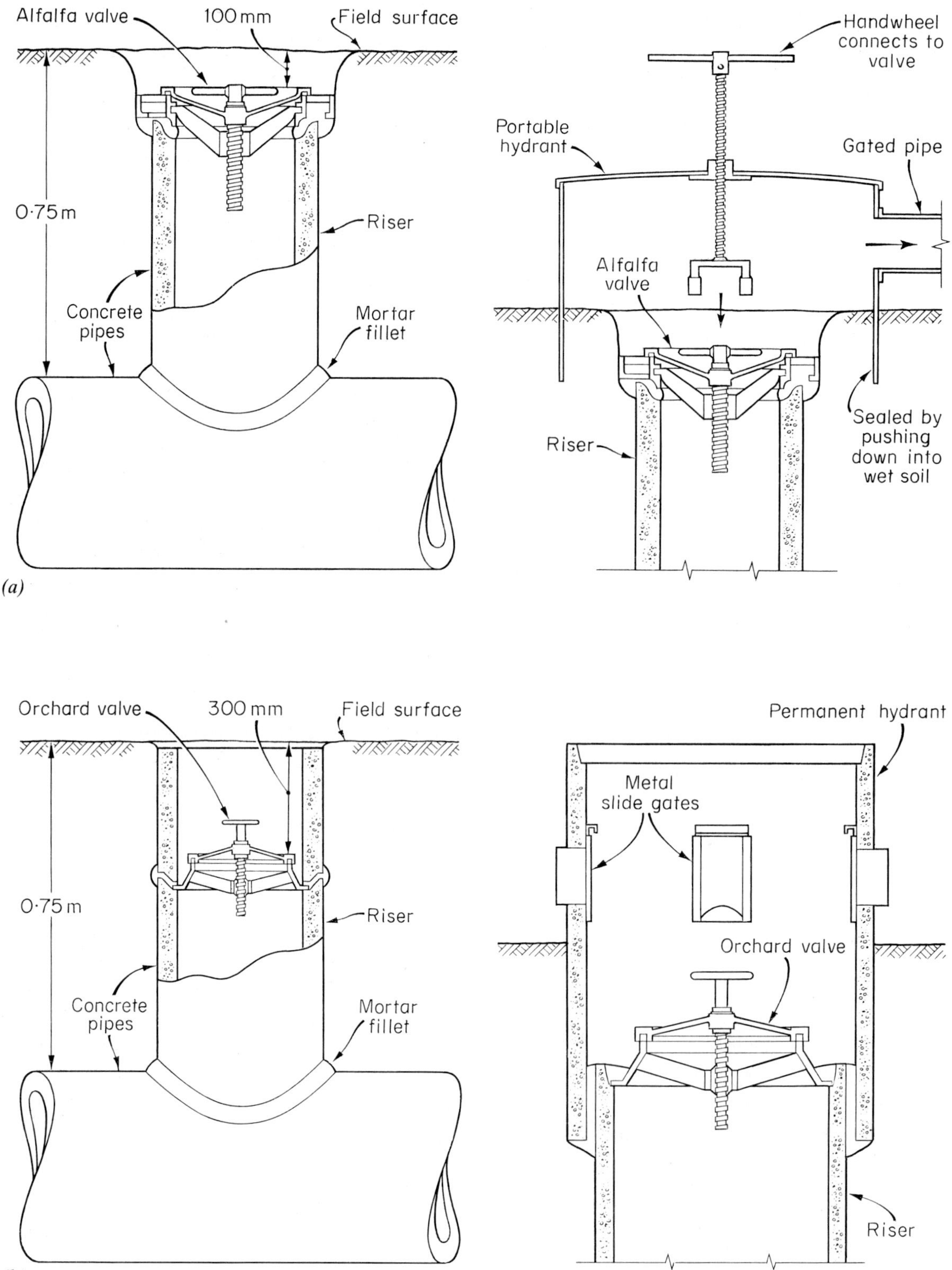

10.5 *Pipe outlets. (a) Alfalfa valve (b) Orchard valve*

Alfalfa valves
These are used for large flows (20-250 ℓ/s) and consist of a steel disc with a screw shaft mounted in a frame on top of the riser pipe (figure 10.5). The size of the valve can vary from 150-500 mm in diameter depending on the discharge. The riser pipe is usually the same diameter as the valve. The discharge from the outlet can be easily controlled by raising or lowering the valve plate.

Orchard valves
These are usually for smaller discharges (1-75 ℓ/s) and sizes vary from 37.5-200 mm in diameter. The valve is set inside the riser and is smaller than the diameter of the pipe (figure 10.5).

Water flowing from both these outlets can cause erosion. For this reason they are usually set about 100 mm below the surrounding ground level. When water is released it forms a pool around the valve which dissipates the energy in the water.

10.5 Hydrants
These are used with alfalfa and orchard valves to improve control over flow from a pipe outlet.

When using gated-pipe a portable hydrant can be used which fits over the pipe outlet (figure 10.5). This is made from aluminium or galvanised steel and it is pushed down into the wet soil around the riser pipe to form a seal. On some hydrants a more effective seal is achieved using clamps which secure the hydrant to the riser pipe (figure 10.2[b]). A screw handle on the hydrant fits over the handle on the valve so that it can be opened.

Another type of hydrant is a more permanent one which is often used with furrow irrigation (figure 10.1 and 10.5[b]). It consists of a concrete pipe placed vertically over the riser with small metal slide gates. The pipe is usually open at the top so that the orchard or alfalfa valve can be adjusted. Flow to individual furrows is adjusted by each slide gate.

10.6 Valves
These are used to control the flow of water in the mainline and laterals. There are two types in common use:

Gate valves

Sluice valves

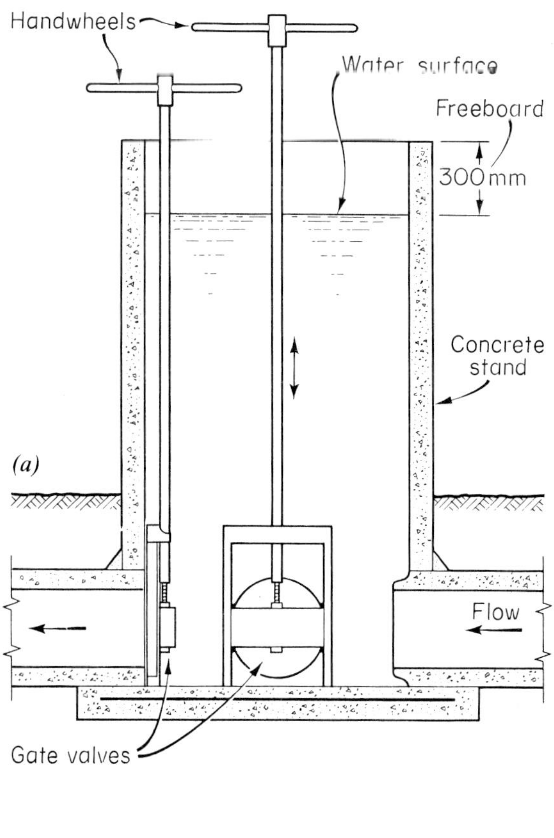

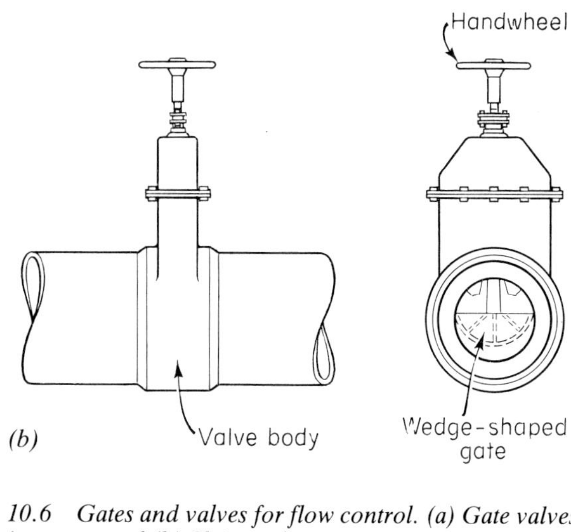

10.6 Gates and valves for flow control. (a) Gate valves in open stand (b) Sluice valve

Gate valves

These are sometimes called *slide-gates*. They are located in open concrete chambers called *stands* (figure 10.6). The gate slides on a vertical steel frame using a hand wheel and screw mechanism. The water level may rise and fall in the stand as the gates are opened and closed and this is a useful measure of the water available in the pipe system. Several gates may be located in one stand from which water is diverted into different pipelines.

Sluice valves

These are sometimes called *line gates* or *gate valves* (figure 10.6). They consist of a valve body connected into the pipelines. A wedge-shaped gate slides inside a groove in the valve body and its movement is controlled by a hand-wheel and a screw shaft. This type of valve is used when operating pressures are too high for open stands.

10.7 Vents

In any pipe system there is always the problem of forcing air out of the pipe when it is being filled with water and allowing air to enter again when draining down. To overcome this problem air vents are used. These are located at each end of a pipeline and at any high spot to let air in and out.

One type of vent is a simple open-topped pipe (figures 10.1[b] and 10.7). This must be tall enough to contain the maximum height to which water will rise under normal operation. Open-topped valve stands will also allow trapped air to escape from a pipeline.

When the operating pressure is too high for an open vent an air valve is used (figure 10.7). This contains a floating ball. When the pipeline is filling with water air escapes round the ball and through an opening in the top of the valve. When it is full water pressure forces the ball against the opening and stops water from escaping. On draining down the pipeline air can re-enter as the ball moves away from the opening. Notice the small air valve located on the left of the hydrant in figure 10.2[b].

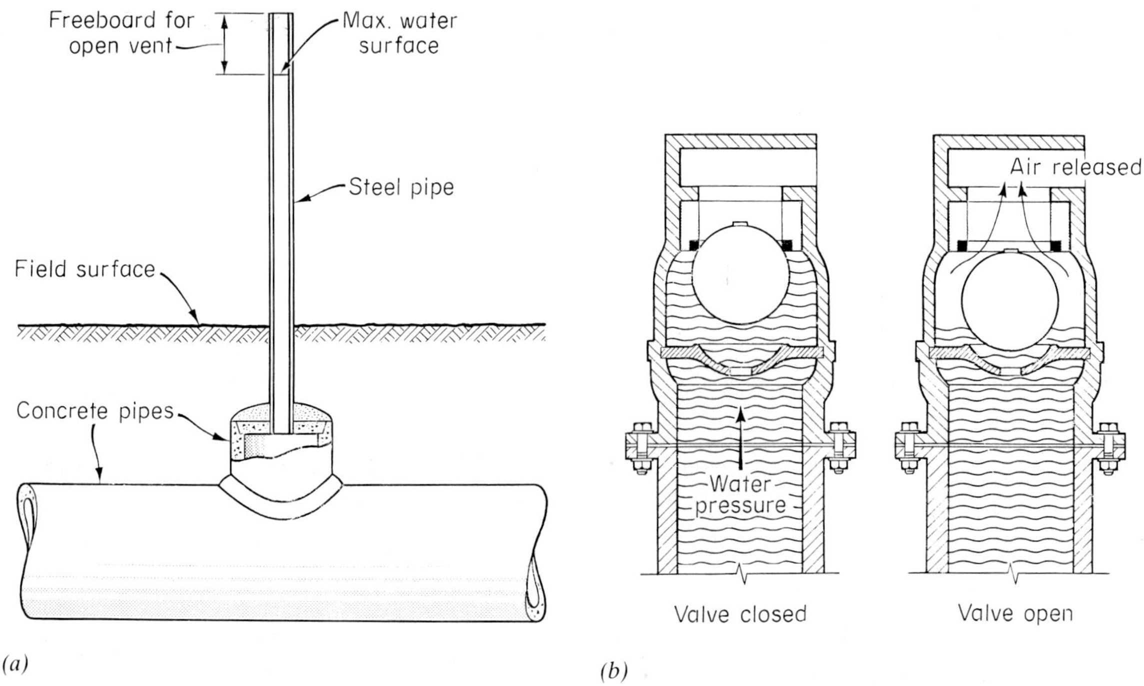

10.7 (a) Air vent (b) Air valve

11

OPERATION

11.1 Introduction

Operating an irrigation system means delivering the right amount of water to farms at the right time. On small schemes consisting only of a few farms this can be easily done. Each irrigator or group of irrigators works together to organise and distribute their own water supply from a river or borehole.

On larger schemes water distribution can be more difficult to organise. The flow must pass through many canals before reaching the farms. Gates must be opened and closed at the right times and the correct flows must be released into each canal. This work is often too much for the irrigators themselves to organise and so people are employed specially to organise the distribution of water from its source, through the canal system, to the irrigation units. Within each unit, the irrigators normally organise water distribution to their farms. Each unit is in fact similar to the small scheme mentioned above.

This seems to be a good way of organising water distribution, but in practice many problems occur and schemes are not always operated properly. Irrigation efficiency is often poor, water supplies are unreliable and some irrigators do not receive their share of the water. There are many reasons for this:

Canal operators may not be properly trained in water management.

There may be a lack of cooperation between canal operators and irrigators.

Irrigators may bribe operators to supply them with more water.

Canal gates may be broken or left open and water is wasted.

There may not be enough water control structures on the canals.

Canal maintenance may be poor. (chapter 12)

Whatever the reason, it is not enough just to design and build a good irrigation system. It must be operated properly if the scheme is to be successful. This means understanding about:

Water demand – water required for irrigation

Water supply – water available for irrigation

Methods of organising water distribution

11.2 Water demand

This is the amount of water required for irrigation. It not only includes the water needed by the crops but also the water losses which can occur on a scheme. These include:

Water losses during irrigation (run-off, deep percolation)

Water losses from seepage and evaporation in the canals

Water distribution losses (Poor organisation).

On some schemes these losses are very high. The largest loss is often in water distribution and is the result of poor organisation.

Water demand varies throughout the season. When crops have just been planted the demand will be low. As the crop grows the demand for water will increase. The irrigation system must be able to cope with the maximum demand.

11.3 Water supply

This is the amount of water available for irrigation. When water is supplied from a reservoir or a borehole the maximum amount available is usually known at the beginning of the irrigation season. When the supply is from a river it is more difficult to decide because the amount available will depend on how much rain falls on the river catchment.

11.4 Matching supply and demand

It is the job of those operating the irrigation system to try and match the supply of water with the demand from the irrigators.

Ideally the water supply should be equal to the demand but often there is a shortage of water. When this happens canal operators and irrigators together must decide how best to distribute the limited supply. This can be done by:

Reducing the area to be irrigated

Changing crops

Changing water management practice

Introducing or increasing water charges.

Reducing the area to be irrigated

This will reduce the demand for water. In many areas it is common practice to grow crops on only 25-30% of the farm area during the dry season. During the wet season when the water demand is less and the supply increases, the cropped area may be increased to 100%.

Changing crops

Crops can be grown which need less water. For example, sorghum can be grown instead of maize or clover instead of lucerne. But irrigators may not wish to change crops for personal and market reasons and so this approach may not be very popular.

Changing water management practices

This can also help to reduce water demand. This includes:

Allocating water to high value crops such as fruit and vegetables.

Reducing irrigation depth (see alternate furrow irrigation page 60).

Extending interval between irrigations.

Introducing a charge for water

This will usually make irrigators use water more carefully. However this can be difficult to do on many schemes because of administrative problems. It will also be undesirable on schemes where farmers are poor and unable to pay for their water.

11.5 Methods of organising water distribution

There are four main methods (figure 11.1):

Continuous flow

Rotational flow

On-demand

Reservoir

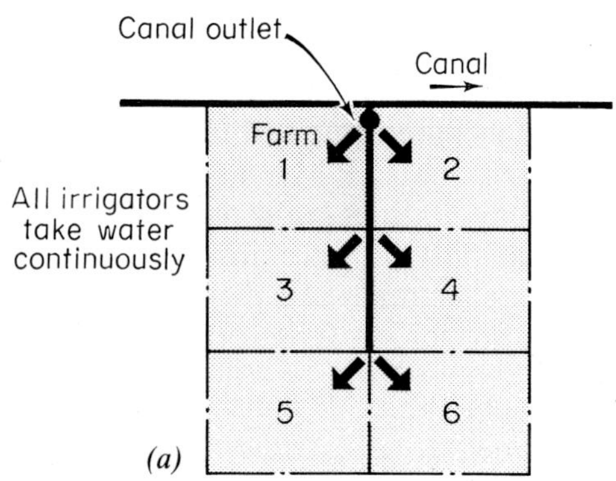

11.1 Methods of distributing water. (a) Continuous supply (b) Rotation (c) On-demand (d) Reservoir (e) Circular weir on a night-storage canal

Continuous flow

This is the simplest and most widely used method but it is also the least efficient. The same discharge flows continuously day and night throughout the irrigation season along the canals and into the irrigation units. Each farm receives a small continuous flow (about 1-5 ℓ/s) usually based on the area of land irrigated.

This method is very simple for the canal operators. The flow in the canals is the same throughout the season and no changes are needed. But it can be difficult for the irrigators. Small continuous flows mean that surface methods are inefficient and irrigators must work long hours, often at night, to irrigate their land. Many irrigators now refuse to work at night in difficult and unpleasant conditions. Water is left to run on the fields at night and a lot can be wasted.

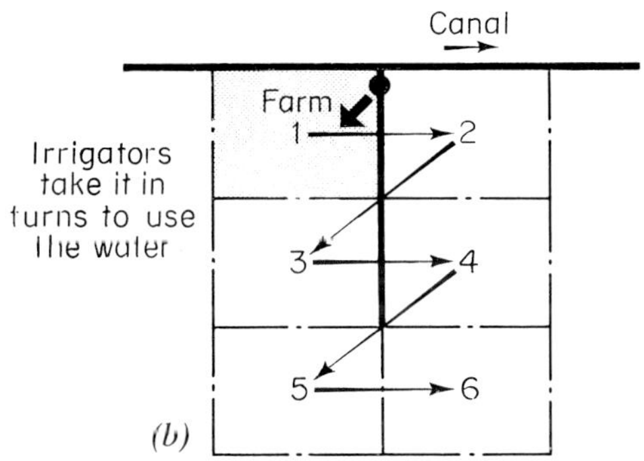

(b)

Rotational flow

This can be more efficient than the continuous method but it requires more control and organisation. Water is supplied continuously in the irrigation system but irrigators within the irrigation unit take it in turns to use the water. Each irrigator receives all the water usually for a fixed time on a fixed day.

Consider the following example. Suppose there are six farms in an irrigation unit using a continuous flow method. Each farm receives 5ℓ/s day and night throughout the irrigation season. The total flow into the irrigation unit is $6 \times 5 = 30$ ℓ/s. Using the rotational method all the flow (30 ℓ/s) would be supplied first to farm 1. When the irrigation was complete the flow would be passed to farm 2 and so on until all six farms were irrigated. The flow would then return to farm 1 to start the rotation again.

This method is used in India where it is called *Warabandi*. Water is rotated on a seven day basis so that irrigators receive water on the same day for the same time each week. Water is shared on a time basis and this is called *time share*. Allowance is made in the time share for the time taken to fill and empty the farm channels. This is important for those irrigators at the far end of the unit when emptying and filling canals can take many hours.

Rotation irrigation means that each person need only irrigate one day rather than every day. When the supply is limited each irrigator can still take his share of the water on the allocated day.

This is the simplest type of rotation. On some of the large irrigation schemes more complex rotations are used. Water is sometimes rotated between distributary canals and the timing of rotations are changed during the season to improve water use efficiency. This can make it more difficult to organise and the canal operators need to be highly skilled.

The success of any rotation method depends on good cooperation between irrigators and canal operators.

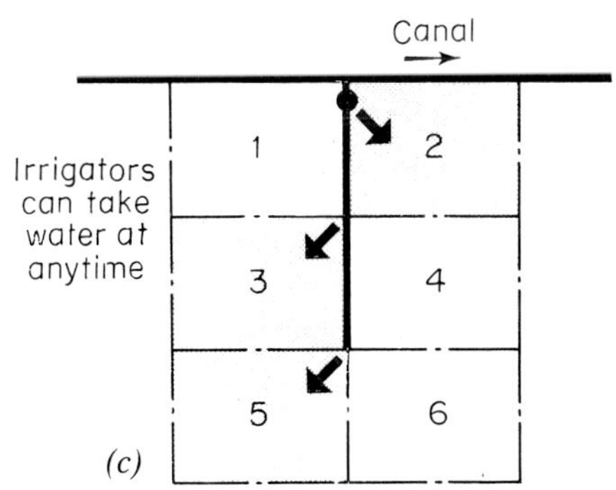

(c)

On-demand

This can be the most efficient method of distribution but it is the most difficult to organise properly. Irrigators can take water from the irrigation system whenever they wish. Usually only a few irrigators will take water at a time and so the demand for water will not be excessive. This is very similar to the supply of domestic water in a town. When someone wants water they turn on a tap and the supply is there. They do not expect to wait for water to come one day each week (as in the rotation method) or to have the tap running slowly all the time (as in the continuous flow method). Although it is very convenient for irrigators this type of system is expensive to construct. Special automatic gates are needed to respond to changes in water demand in the canals. The system does not work well when too many irrigators decide to irrigate at the same time. Cooperation between irrigators is essential to prevent the irrigation system from being over-loaded.

Reservoir

This combines the easy operation of the continuous flow method with the convenience of the on-demand method for the irrigators. Reservoirs are built along each distributary canal and are filled with water continuously day and night. Irrigators then take the stored water when it is convenient to them.

The reservoirs are sometimes called *night-storage reservoirs* because they store water at night so that irrigators need only irrigate during the day. On some schemes water is stored in large distributary canals. These are called *night-storage canals*. Special circular weir structures are used to control the storage level in each section of canal (figure 11.1).

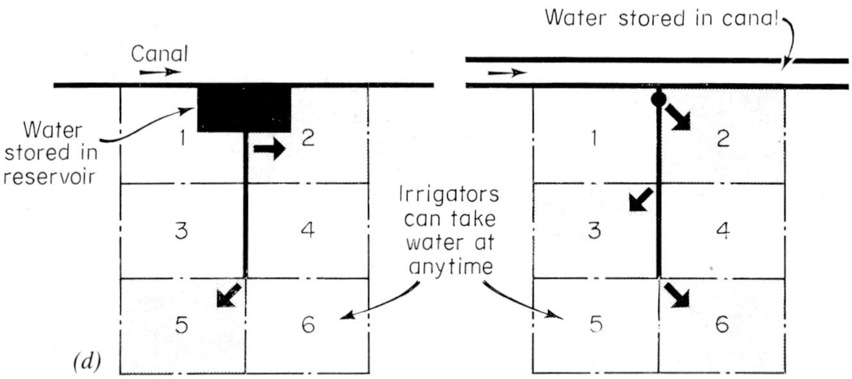

(d)

(e)

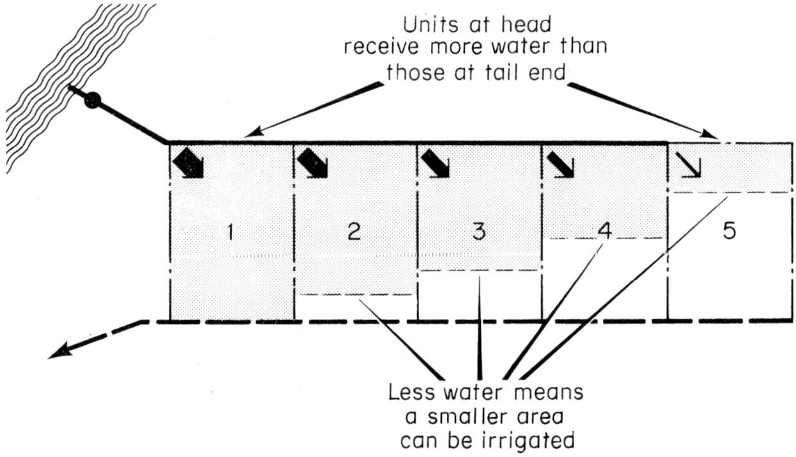

11.2 Common problem – unequal distribution of water.

11.6 Common problems

Many problems occur when operating irrigation schemes the following are some of the most common.

Poor water distribution

Some irrigators often get more than their share of water and some get less (figure 11.2). There are many reasons for this. It may be due to control gates not being set properly or irrigators blocking the canal and diverting more water into their farms. It may be due to some irrigators cheating or bribing gate operators. Whatever the reason those at the tail end of the canal usually suffer more than those near the head. This is often because those irrigators nearest to the water supply will be tempted to take as much water as they need without thinking of others further down the system.

This problem can be overcome by installing special gates and locks to try and stop people taking too much water but it will not stop corruption. This is a social problem and not a technical one. It can only be solved by the authorities, the canal operators and the irrigators themselves.

Low water demand

This can occur when a scheme first comes into operation and during the early stages of crop growth. Irrigation schemes are designed to supply the peak water demand and these early periods of low demand can cause operating problems.

As an example, suppose a distributary canal normally flows at 60 ℓ/s supplying 5 units with 12 ℓ/s each (figure 11.3). If the demand at the beginning of the irrigation season is only 50% of this (30 ℓ/s) then all the

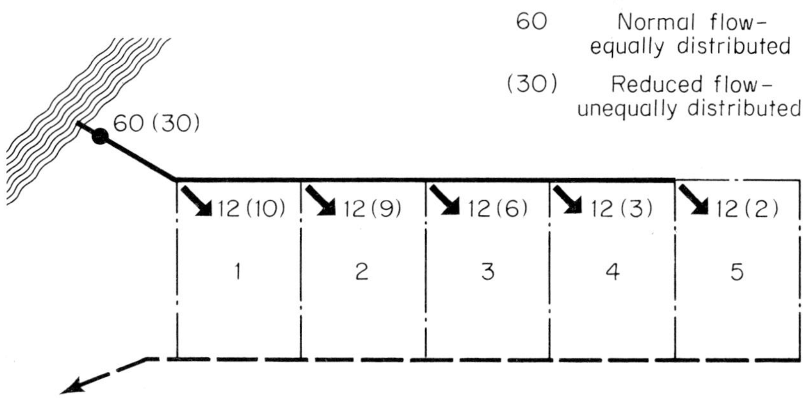

11.3 Common problem – low water demand.

outlets along the canal could expect to receive 50% of their normal flow (6 ℓ/s). This is called *proportional distribution*. However, on many schemes the reduced flow is not proportionally distributed. Canal outlets near the head of the canal usually get more water than those at the tail end.

There are two ways of overcoming this problem. When canal outlets are installed with adjustable gates the gatemen can adjust each outlet when the flow in the canal changes. But many outlets do not have such gates and can only be fixed open or closed. Also the work can be very demanding when there are a lot of gates to adjust and the canal flows are frequently changing.

The second way can simplify operation. The distributary canal is operated at its normal flow and the operating time period is reduced. So in the above example the canal would be operated at 60 ℓ/s for half the time rather than 30 ℓ/s all the time. For the rest of the time period the canal would be closed. Adjusting the operating time period instead of the discharge means that a wide range of operating conditions can be achieved with very little adjustment needed to the canal gates.

Limited water supply

When the supply is less than the demand some canal operators overcome this problem by reducing the discharge in the canals. This can produce similar problems to those just described for periods of low demand. The solutions are also similar.

Operating canals at normal flow for part of the time can also have the advantage of reducing seepage losses. This saves water at a time when it is most needed.

Unreliable supply

This occurs on many schemes and is one of the most difficult problems to solve. The irrigator is unable to plan his cropping properly because he is not sure about the amount of water he will receive and when it will be delivered.

There are many reasons for this. It may be due to breakdowns in the irrigation system because of poor maintenance but it is often a social problem rather than a technical one. This is the lack of cooperation and understanding between the canal operators who supply the water and the irrigators that use it.

Technically, improved maintenance will help to improve the reliability of supply. Using the reservoir method of distribution can also help to overcome this problem. The irrigator can see how much water is available in the reservoir and so is able to plan ahead.

The social problems of communication between the canal operators and the irrigators can be solved partly by training but the real solution lies within the community itself.

Filling and emptying canals

When rotation irrigation is used or when canals are only used part-time, time must be allowed for filling and emptying the canals. It can take many hours to fill a long canal with water before irrigation can begin. The time depends on the discharge and the length and size of the canal.

As an example suppose a farm channel 1 kilometre long with a cross sectional area of 0.75 m^2 is to be filled with water before irrigation can begin (figure 11.4). How long will this take when the discharge is 30 ℓ/s (0.03 m^3/s)?

$$\text{Volume of water required } (m^3) = \text{length of channel (m)} \times \text{cross sectional area } (m^2)$$

$$= 1000 \times 0.75$$

$$= 750\ m^3$$

$$\text{Time to fill channel (h)} = \frac{\text{volume } (m^3)}{\text{discharge } (m^3/s) \times 3600}$$

$$= \frac{750}{0.03 \times 3600}$$

$$= \underline{7\text{ hours}}$$

Always make sure the units of measurement are correct when doing calculations of this kind.

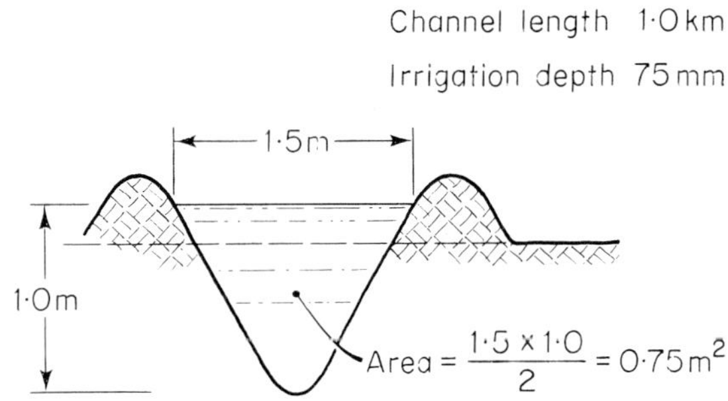

11.4 Calculating filling/emptying time for a farm channel.

Equally there can be a lot of water left in a canal at the end of an irrigation. This water need not be wasted. It should be possible to drain down the canal and use most of the water to irrigate units at the lower end of the system. It also means that flow into the farm channel should be stopped long before field irrigation is completed to avoid wasting water.

Continuing the previous example, if the irrigation depth is 75 mm, what area of land could be irrigated from draining down the farm channel? At what time can the flow into the farm channel be stopped?

It will not be possible to take all the water from the farm channel unless it is pumped. Assuming that 80% of the water can be used (80% of 750 m^3 = 600 m^3)

$$\text{Land area (ha)} = \frac{\text{volume of water } (m^3)}{10 \times \text{irrigation depth (mm)}}$$

$$= \frac{600}{10 \times 75}$$

$$= \underline{0.8 \text{ ha}}$$

$$\text{Time to empty channel (h)} = \frac{\text{volume of water } (m^3)}{\text{discharge } (m^3/s) \times 3600}$$

$$= \frac{600}{0.03 \times 3600}$$

$$= \underline{5.5 \text{ hours}}$$

Flow into channel should be stopped 5.5 hours before field irrigation is complete to avoid wasting water.

11.7 Manpower

The staff needed to operate the irrigation system are part of the overall management needed to organise and run an irrigation scheme (figure 11.5).

The number of people and the skills needed depends on the size of the scheme, the method of water distribution and the number and size of the farms. The names given to each group differ greatly from one country to another but their jobs are very similar. The following are typical job descriptions and areas of responsibility.

Gate operators

These people are responsible for adjusting the gates on head and cross regulators and canal outlets. They normally work closely with the water guards who pass on instructions to them about when to open or close canals or to change the flow. Gate operators may also carry out routine maintenance on the gates such as greasing and removing trash which may cause blockage. Any serious repairs would normally be carried out by skilled maintenance staff.

Pump operators

These are only needed if water is pumped from a river or borehole. Their responsibilities would include:

Starting and stopping pumps as required

Maintaining pumps and power units

Maintaining fuel supplies, lubricants and spare parts.

They would normally receive their operating instructions from the water guard or directly from the irrigation engineer. Most people employed in this capacity may not have had any formal education but received some practical training for their job.

Water guards

This is a most important group of people because they are the main link between the engineers who operate the system and the irrigators in the field. Sometimes they are

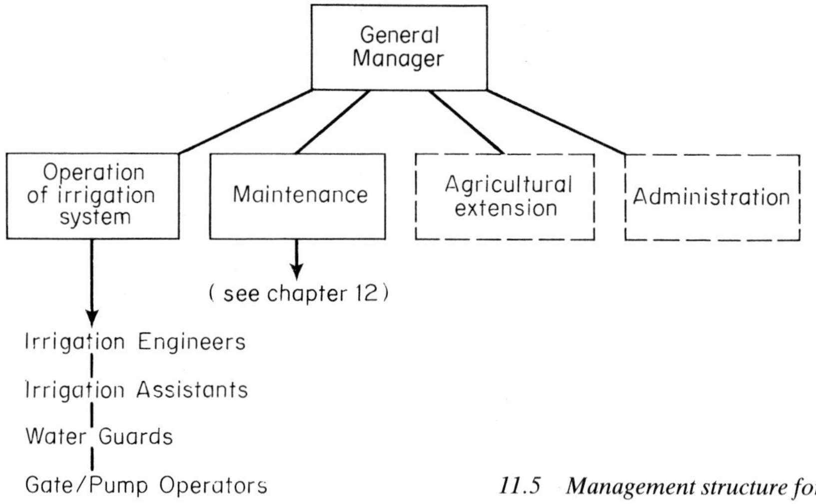

11.5 Management structure for a large scheme.

called *irrigation overseers, ditch riders* or *water bailiffs*. Their work involves:

Helping irrigators with water management problems

Checking water use efficiency on farms.

Passing instructions to gate operators

Collecting requests for water from irrigators

Reporting problems on land preparation and canal maintenance.

Each water guard would normally be responsible to an Irrigation Assistant and look after 400 ha or more. This depends on the type of system and the size and number of farms. He would need at least primary school education and 1-2 years practical training.

Irrigation assistant
These are sometimes called *superintendents, master water guards* or *head water bailiffs*. Their task is to coordinate the work of a number of water guards. This may include:

Collecting information from water guards on water requirements, maintenance needs and land preparation and passing this to the irrigation engineer.

Passing instructions on water allocations from the irrigation engineer to the water guards.

Organising training courses for water guards, gate operators and irrigators.

An Irrigation Assistant would be responsible to an Irrigation Engineer and looks after four or more water guards covering about 1000 ha. This depends on the system and the size and number of farms. He would need at least secondary school education followed by 2-4 years in a college studying for a diploma in irrigation.

Irrigation engineer
An engineer will usually be responsible for the operation of the whole irrigation system. His work would include:

Supervising irrigation assistants

Preparing irrigation plans

Working closely with agricultural staff to plan the supply and demand for water

Reporting all maintenance needs to the maintenance department

Reporting to the general manager on the irrigation system and its future needs.

An irrigation engineer could be responsible for up to 5000 ha or more. He would normally have a university degree in engineering with special training in irrigation.

EVALUATION

When operating an irrigation system it is important to know if it is working properly.

Is the right amount of water being delivered to all parts of the scheme?

Is water being used efficiently?

Are there any signs of longer term problems such as salinity or soil erosion?

To answer these questions an evaluation of the irrigation system and farm irrigation practices can be carried out. On a small scheme this may be done by the irrigator with help from the local irrigation office or extension service. On a large scheme it may be carried out by the canal operating staff in cooperation with the irrigators.

11.8 Irrigation system

A very simple but useful evaluation of an irrigation system can be made by careful observation and asking questions. The water guard responsible for each distributary canal can regularly check the following (figure 11.6);

Ask irrigators along the canal if they are getting enough water. Those who are short of water will no doubt complain. Beware of those irrigators who always complain even when they get more than enough!

Check the drainage system. If there is water flowing in the drains, find out where it is coming from. It may be rainfall causing surface run-off or it may be poor use of irrigation water. It may also be run-off from furrow irrigation. Is it coming from all the units or from just one or two? Look for

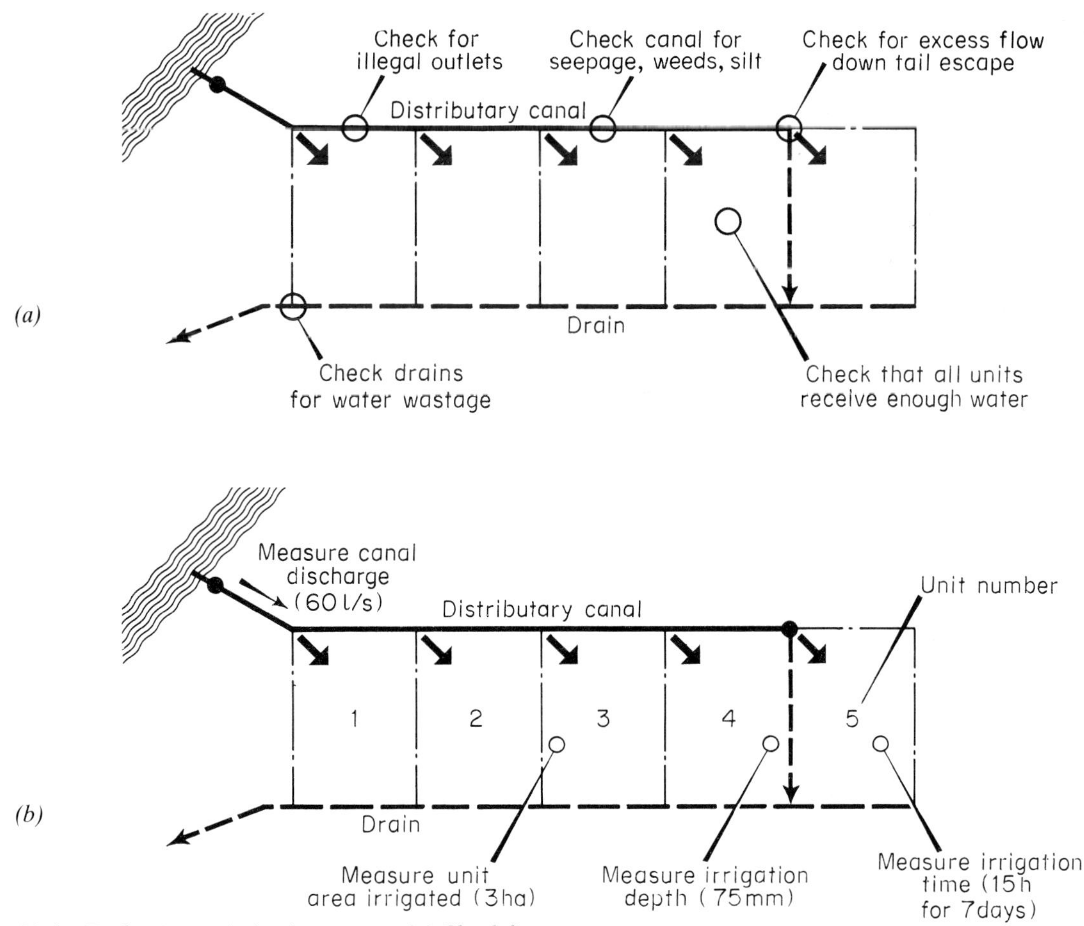

11.6 Evaluating an irrigation system. (a) Check list (b) Measurements (c) Installing weir to measure distributary discharge

ways of reducing these losses.

In flooded rice schemes check to see if there is the right depth of water on the fields.

Check along the distributary canal to see that all the canal outlets are working properly. Sometimes irrigators will install extra outlets in the canal to get more water. Make sure there are no such illegal outlets.

Check the end of the distributary canal. If water is flowing down the tail escape, there is too much water entering the canal.

Make sure irrigators are using all the water they receive. If they arrive late in the morning or leave early in the evening a lot of water can be wasted.

Check along the canal for signs of seepage and leakage at structures, weeds and siltation.

(c)

When there are flow measurement structures on the distributary canal a more detailed evaluation can be done. This involves comparing the water required for irrigation and the water supplied in the canal.

Calculate the volume of water required on each unit. To do this the irrigated area and the irrigation depth must be known.

Volume required for unit (m^3) = 10 × irrigated area (ha) × irrigation depth (mm)

Calculate total volume of water required in the distributary canal.

Volume required from canal (m^3) = Volume for unit 1 + Volume for unit 2 + ...

Calculate volume of water supplied to the distributary canal. To do this measure the discharge at the head of the canal, the number of days and the number of hours each day that water flows.

Volume supplied by canal (m^3) = 3.6 × canal discharge (ℓ/s) × number of hours (h) × number of days (d)

Compare volume of water supplied with the volume required and calculate the water loss and the efficiency of the canal.

Water loss (m^3) = volume supplied − volume required

$$\text{Canal irrigation efficiency (\%)} = \frac{\text{volume required}}{\text{Volume supplied}} \times 100$$

Example

The following measurements were made to evaluate water loss and canal irrigation efficiency on a distributary canal (figure 11.6).

Unit area	3 ha
Number of units	5
Irrigation depth	75 mm
Discharge at head of canal	60 ℓ/s
Irrigation time	15 h each day for 7 days

Using the above procedure:

Volume required for unit = 10 × 3 × 75 = 2250 m^3

Volume required from canal = 2250 × 5 = 11250 m^3

Volume supplied by canal = 3.6 × 60 × 15 × 7 = 22680 m^3

Water loss = 22680 − 11250 = <u>11430 m^3</u>

$$\text{Canal irrigation efficiency} = \frac{11250}{22680} \times 100$$

$$= \underline{50\%}$$

A lot of water is clearly being lost and efficiency is low. The irrigation system should be inspected carefully for signs of water wastage.

Sometimes this calculation can be misleading. For example, when there is a large demand for irrigation water and only a small amount is supplied the calculation may show a good efficiency when the irrigation may be very poor. The calculation will not show the severe water shortage.

11.9 Irrigation method

When water losses occur in the unit an evaluation of the irrigation practice is needed. Many of the most common faults in basin, border and furrow irrigation have already been described in other chapters. Careful observation of these irrigation practices may show where water is being wasted. If the losses are not so obvious a more detailed evaluation can be carried out. The following is related to basin irrigation as this is the most common irrigation method. Similar tests can be done for borders and furrows.

The following equipment will be needed:

Siphon pipes (those normally used by the irrigator)

Rectangular weir plate

Steel measuring tape (30 m)

Soil probe

Bucket (5 ℓ capacity)

Wooden posts

Hammer

Stop watch

Infiltrometer

Locate a typical basin on the farm in terms of size, crops and soils that is ready for irrigating.

Set out the wooden posts at 10 m intervals across the basin.

Measure irrigation depth.

Carry out an infiltration test in the basin. (section 2.10)

Start the irrigation and note the time. Use the same flow as the irrigator normally uses.

Measure the stream size used for irrigation. This can be done by installing a small rectangular weir in the canal (figure 2.4). If siphon tubes are used the flow can be measured using the volume method (figure 2.7).

Record the time taken for water to advance across the basin at each wooden post.

Record the time when the basin has been filled with water and stop the inflow.

Record the time taken for all the water to infiltrate (recession). Check to see that water infiltrates evenly over the whole basin. This will show how well the basin has been levelled.

Check the adequacy of irrigation a day later using a soil probe. Push the probe into the soil in different parts of the basin and record the depth of wetting. The probe can be pushed easily through the wet soil. It will be much more difficult to push when it reaches the drier soil below.

The following example shows how the collected data can be used.

Example

In the previous example a lot of water was being lost somewhere in the scheme. To try and locate these losses an evaluation was carried out on a typical basin in one of the units. The data collected is shown in Table 11.1 and 11.2 and in figure 11.7.

The irrigation depth to be applied was 75 mm. The irrigation started at 10.00 hours and the time taken for the water to advance across the basin was recorded in column (2). The elapsed time in column (3) is the number of minutes since the start of the test. In column (4) the time the recession occurred was recorded. The recession time is the same over most of the basin because the land is flat.

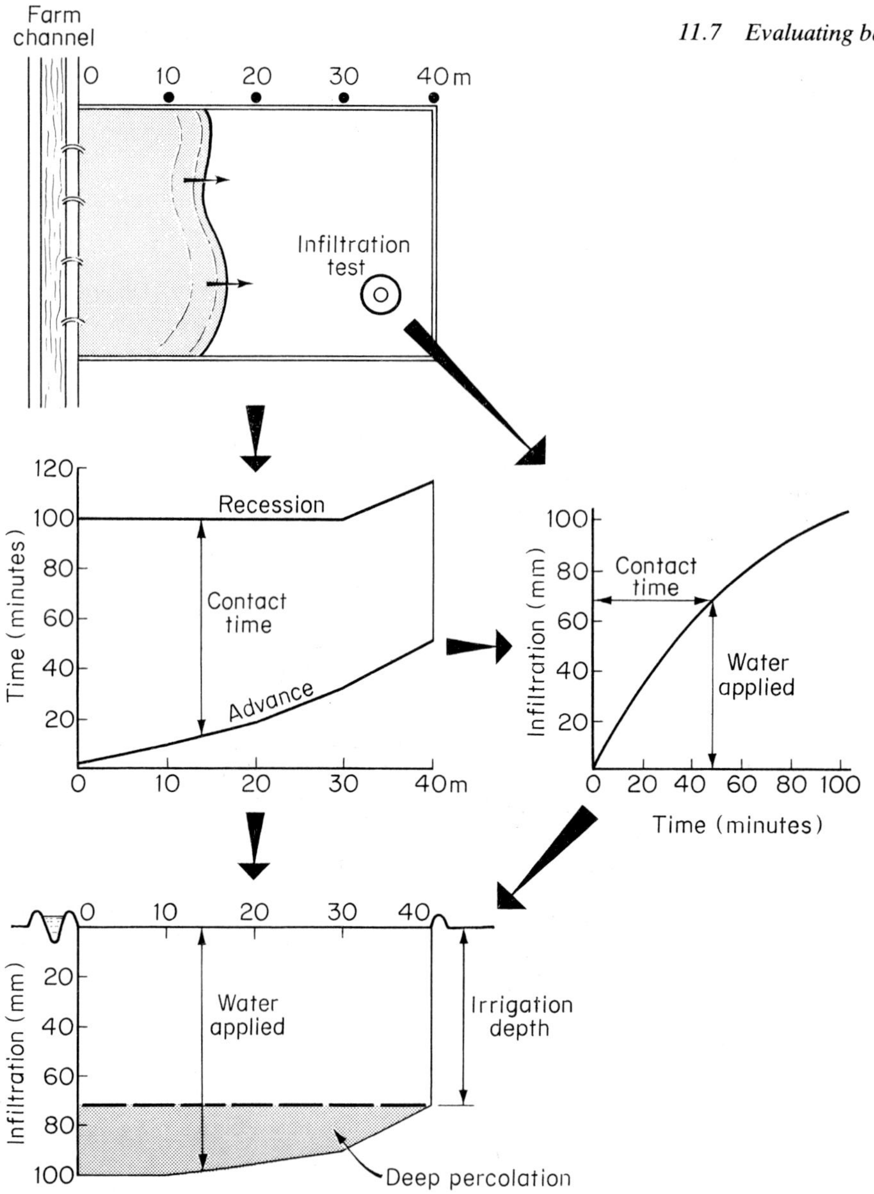

11.7 Evaluating basin irrigation.

Table 11.1 Basin irrigation evaluation data

DISTANCE (m) (1)	ADVANCE TIME clock (2)	ADVANCE TIME elapsed (minutes) (3)	RECESSION TIME clock (4)	RECESSION TIME elapsed (minutes) (5)	CONTACT TIME (5)-(3) (minutes) (6)	WATER APPLIED (mm) (7)
0	10.00	0	11.39	99	99	103
10	10.07	7	11.39	99	92	101
20	10.17	17	11.39	99	82	95
30	10.31	31	11.39	99	68	90
40	10.49	49	11.50	110	61	77
					Average	93.2

Irrigation depth = 75 mm

However, at the far end it takes a little longer showing there is a low spot where too much water has ponded. Column (5) is the recession time elapsed since the start of the test. The contact time in column (6) is found by subtracting (3) from (5). This is the contact time at each place across the basin. In column (7) is the water applied at each place across the basin. It is obtained using the contact time and the infiltration data (table 11.2).

Table 11.2 Infiltration test results

TIME (minutes)	*WATER INFILTRATED* (*mm*)
5	15
10	25
20	40
30	52
50	74
70	90
100	103

The results show that a little too much water is applied at the far end of the basin (77 mm) but far too much is applied near the farm channel (103 mm). The average depth applied is 93.2 mm (table 11.1) compared with the required irrigation depth of 75 mm.

Using these values to evaluate the farm:

Volume required for unit (m^3) $= 10 \times$ irrigated area (ha) $\times$ irrigation depth (mm)

$= 10 \times 3 \times 75$

$= 2250\ m^3$

Volume supplied to unit (m^3) $= 10 \times$ irrigated area (ha) $\times$ average depth applied (mm)

$= 10 \times 3 \times 93.2$

$= 2796\ m^3$

Water loss in unit $= 2796 - 2250$

$= \underline{546\ m^3}$

Irrigation efficiency $= \frac{2250}{2796} \times 100$

$= \underline{80.5\%}$

This is a good efficiency for basin irrigation but it can be improved. Firstly, the far end of the basin should be levelled. Secondly, the advance should be much faster to reduce deep percolation. The contact time is 50 minutes for an irrigation depth of 75 mm. Using the quarter time rule the advance time should be 50/4 = 12.5 minutes. This means using a larger stream size so that water will flow faster across the basin.

The deep percolation loss in the unit has been calculated at 546 m^3. If there are similar losses on all 5 units the total loss is 5 × 546 = 2730 m^3. This is small in comparison to the total water loss of 11430 m^3. This means that the main losses occur in the irrigation system and not in the field irrigation.

12

MAINTENANCE

12.1 Introduction

This is the routine work which must be done to keep an irrigation scheme working properly. It involves cleaning and repairing canals and drains, repairing structures and canal linings and cleaning and lubricating gates.

Unfortunately on many schemes, maintenance is neglected. Particularly the small repair jobs such as filling in rain gulleys in embankments. When these are left they grow rapidly until it becomes a major job to fill them. Canals and drains often become choked with weeds and silt and gates do not work because they are broken or rusted through lack of attention.

There are many reasons why maintenance work is not done. Sometimes irrigators and canal operators do not understand its importance and so have no interest. They often prefer to make repairs when something breaks down rather than making sure the canals keep running properly. On many larger irrigation schemes very little money is made available for maintenance and so there is often not enough equipment, men and materials to do the work. Whatever the reasons, if the system is not maintained properly then it cannot be expected to work well.

Maintenance requirements will differ from one scheme to another depending on the size and the irrigation practice. The following are some of the most common problems and ways of solving them.

OPEN CHANNELS

12.2 Unlined canals

There are four main problems:

Silting

Weed growth

Burrowing animals

Erosion of banks

Silting

This is a very common problem and is mainly due to silty water being taken from a river. The silt settles on the bed of the canal and reduces the flow capacity. Siltation can be a problem on new canal systems when they are operating at very low flows in the early stages of development. Some canals can become completely blocked when the silt content of the water is very high.

Silt can be removed by hand but this is usually limited to watercourses and farm channels. To make sure that the correct shape of channel is excavated a profile board of the correct shape can be used. This is inserted into the channel at regular intervals during excavation. The labourers can then see whether or not more excavation is needed. Canals must be emptied and allowed to dry out for several days before excavation begins. This not only makes the work easier but it also reduces the risk of infection from water-borne diseases (see chapter 13).

(a)

12.1 *Clearing silt from canals. (a) Dragline (b) Operating details*

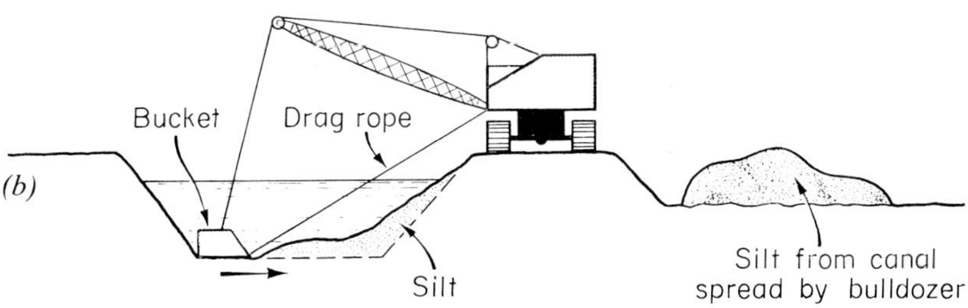

(b)

The most common machines used for clearing silt are draglines and hydraulic excavators (figure 12.1). These machines work from the canal bank top and are usually crawler mounted for easy working in muddy conditions. They have a reach of 10-15 m. Small canals can be cleaned by one machine working from one of the bank tops. Larger canals usually require two machines working on opposite banks.

Although the excavators can work while the canal is flowing it is difficult to see how well the work is proceeding. Silt may also be carried further downstream by the flow. Excavation is best done at the end of the irrigation season when the canal can be drained down. The silt is dumped outside the canal, left to dry and spread by bulldozer. Up to 300 m of canal can be cleared each day depending on the working conditions.

(a)

12.2 *Clearing weeds from canals. (a) Dense weed growth can greatly reduce water flow (b) Hand clearance (c) Hydraulic excavator with weed cutting bucket*

Weed growth

Silting encourages weed growth. In some cases the weeds become so dense they almost stop the flow of water (figure 12.2).

There are many types of water weed. Some grow on the canal banks close to the water. Others grow in the water and are either submerged or float on the surface.

Water weeds are usually removed by cutting or by excavation when silt is being removed. They can be cut manually or by machine. Hand tools such as scythes and rakes can be used (figure 12.2). The right tools will mean that the work can be done quickly but large numbers of labourers are needed. One man can usually clear from 50-150 m of farm channel each day. On large canals excavators can be used fitted with special weed cutting buckets. Weeds are best cut near the base of the stem so that the roots are left undisturbed. Regular cutting in this way will weaken and kill the plants.

Chemicals are available which can kill aquatic weeds. However, they can be expensive and great care is needed when they are used. Some of the chemicals are harmful to people and animals and also to crops. They should not be used in places where people regularly use the canals for drinking and bathing.

Water weeds are not always a nuisance. Reeds are often cut and used as roofing material for houses.

Burrowing animals

Leakage through canal banks can be caused by small burrowing crabs and rats or by rotting of plants not removed during construction.

Small holes can be plugged with soil using spades and small hand compactors. Larger holes may need machines to excavate out the damaged section of bank and replace it with new compacted soil.

(b)

(c)

Erosion of banks
Canal banks can be eroded by rainfall, wave action (figure 12.3), animals using the canals for drinking and heavy vehicles using them as roads. The worst damage is usually done in wet weather.

Gullying from rainfall is probably the most common erosion problem. Repairs are best done at the end of the rainy season when holes can be filled with soil and compacted either by hand or by machine. Where gullying leads to excessive seepage and a risk of breaching the bank immediate repairs are needed.

Erosion can be greatly reduced by growing grass on the banks. This provides protection for the soil and also slows down the water running off the side slopes. Cattle damage can be reduced by constructing special places for animals to drink from canals. Vehicle damage can be reduced by stopping vehicles using the canal banks for access during wet weather.

12.3 Lined canals
These are usually built to reduce maintenance problems. They will do so providing they have been built properly.

After some time it may be necessary to replace sections of cracked concrete lining and damaged joints (figure 12.4). This can be expensive and requires skilled masons to do the work well. Normally silt and weed growth are not a problem as settling basins and silt traps are used to get rid of silt. If silting does occur then weeds may also grow. This can only be removed by hand because machines would damage the lining.

12.4 Drains
These are at the end of a scheme and tend to be the most neglected.

Silt is not usually a problem because most of the flow is from rainfall and excess irrigation water. However weed growth can be very heavy. For practical purposes maintenance of open drains is very similar to that of unlined canals. When clearing drains it is best to work upstream from the outfall so that water can flow easily out of the scheme.

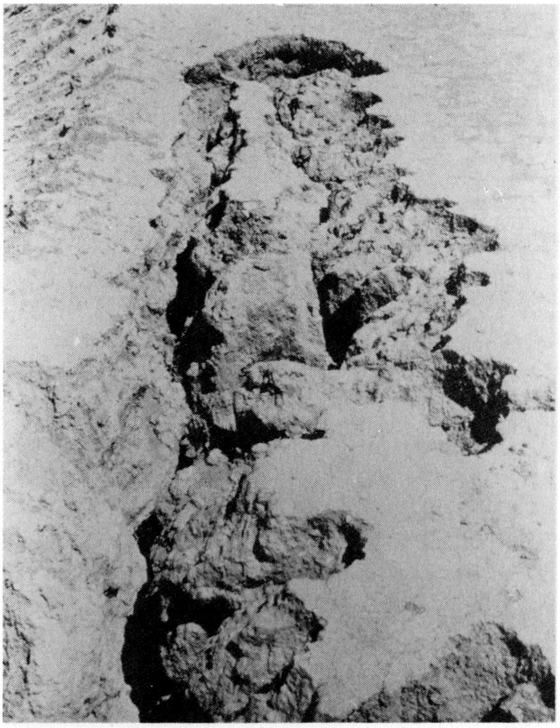
(a)

(b)

12.3 Erosion of canal banks. (a) Gully formed by rainfall (b) Bank collapse from wave action

12.4 Damaged concrete canal lining.

(a)

12.5 *Canal breaches. (a) Effect of breach on a branch canal regulator (b) Repairing a breach in a watercourse*

(b)

12.5 Breaches

Breaches are breaks which occur in channel banks causing water to spill from the channel in an uncontrolled manner.

In canals they occur when banks have not been properly compacted during construction, when they have been eroded and not properly maintained; and when water levels rise because of mis-use and water flows over the bank top.

Breaches on large canals can be very dangerous. People can be drowned and a great deal of damage done by the sudden release of large flows of water (figure 12.5). Repairs are often done in an emergency using a bulldozer to push up soil to close the gap in the bank. This is a bad practice because it leaves the bank very weak. The soil is not compacted and it is sitting on a loose muddy foundation. The best method is to drain down the canal; remove all the mud and debris from the damaged section; and re-build the bank with new compacted soil. Care is needed to compact the soil where it ties into the existing bank.

Breaches in small channels such as farm channels and water courses can be repaired by hand. If the flow is not too high (50 ℓ/s or less) the breach can be repaired while the water is flowing (figure 12.5). For larger flows it will be necessary to drain down the channel.

In drains, breaches are usually deliberate. Irrigators sometimes open the protecting bank around drains to release flood water or excess irrigation water when the small drain inlets cannot cope with the flow. This can cause severe gully erosion and the material washed down can block water flow and cause flooding. Irrigators should not be allowed to continue such damaging practices.

STRUCTURES

Canal and drain structures must also be carefully maintained for the system to work properly.

12.6 Water control gates

Most irrigation schemes use moving gates to control the distribution of water. These need regular attention if they are to continue to function.

Many of the more modern gates are supplied with handbooks by the manufacturer giving details of greasing, oiling, general cleaning and checking movement over full range of travel. Water seals are usually inspected when the canal is empty and they are replaced if worn or perished.

Gates are usually made of steel and this requires regular painting to avoid rust.

12.7 Concrete, blockwork and stone

This type of construction does not usually need maintenance except for the removal of weeds and silt which may obstruct the water flow. Sometimes cracks occur in concrete and settlement takes place. If this affects the structure or endangers the canal in any way major repairs will be needed.

(a)

12.6 Hydraulic structure protection problems. (a) Stone pitching removed from a drain inlet to build a crossing (b) Farm structure failed due to poor downstream protection

12.8 Protection

Most hydraulic structures are protected from erosion by stone or block pitching. This too needs careful attention. Sometimes stones are removed by people and used for other purposes. The example shown (figure 12.6) indicates a need for a small bridge to be built across this drain channel. Another example (figure 12.6) shows how a small farm structure has failed because of poorly maintained downstream protection.

(b)

ORGANISING MAINTENANCE

12.9 Types of maintenance

Maintenance is best done to a regular programme and time must be allowed each year for work to be carried out.

There are three types of maintenance:

Routine

Annual

Emergency

Routine

This refers to the simple jobs which are done daily such as cleaning and greasing of gates, keeping structures clear of floating debris and filling in small rain gulleys on canal and drain embankments. The general condition of the whole scheme depends on irrigators and canal operators attending to these small but important tasks.

Annual
This is usually cleaning and re-shaping of canals and drains, repairs to structures and grading embankments to improve access roads. On large schemes it is important to plan a programme of work so that enough men, equipment and materials can be made available.

Canals are usually cleaned during the period when water is not needed for irrigation and part or all of the canal system can be closed. Some may require cleaning several times each year, others may only need cleaning every two or three years. It depends on the canals, how they are used and their location.

Drains cannot be closed for maintenance but the work can be done when the flows are low. This is usually in the dry season when all the water is used for irrigation and there is no surplus. The channels are then clean and ready for the rainy season. Drains are usually cleaned starting at the outfall and working back up through the drainage system.

Road maintenance is also important, particularly the inspection roads along the canals and drains. This will improve access for other maintenance activities.

Emergency
On any irrigation scheme problems can occur unexpectedly such as a canal bank breaching or gates jamming. It is important to be able to handle these problems quickly so that further damage can be avoided. Normally machines and labour would be diverted from routine maintenance to handle an emergency.

Inspection
An important part of maintenance work is inspection. This is the regular examination of canals, drains and structures and the reporting of any problems to the group responsible for organising maintenance. Day to day inspection of farm channels and water courses would normally be done by the irrigators. Distributary, branch and main canals would be inspected by gate operators and irrigation system operating staff.

On large schemes an inspection of the whole system is usually made each year by an engineer in charge of maintenance to make sure that work is being done properly and that his men and machines are being used efficiently.

12.10 Manpower and equipment
The staff and equipment needed to maintain an irrigation scheme will depend on many factors, including size and the type of irrigation and drainage system used. Normally a maintenance department would be responsible for maintaining its own machinery as well as the irrigation and drainage system. Typical equipment normally used for maintenance is listed in table 12.1. In table 12.2 is a list of staff and skills normally required.

Table 12.1 Typical maintenance equipment for a large scheme

Type		*Comments*
Draglines/hydraulic excavators	—	canal/drain cleaning
Bulldozers	—	general earthmoving
Motor graders	—	grading roads
Sheep-foot roller	—	for compacting canal banks
Lorries	—	for general earth moving
Low-loader	—	transporting heavy machinery
Pick-up trucks	—	for maintenance staff and small equipment
Concrete mixers	⎤	
Vibrators	⎥	
Water tankers	⎬	concrete work
Pumps	⎥	
Compressor and tools	⎦	
Welding equipment	⎤	
Lifting tackle	⎬	gate maintenance
Small crane	⎦	
Hand tools	—	general repairs

Table 12.2 Typical maintenance staff for a large scheme

Staff	*Comments*
Engineers – civil – mechanical – electrical Office staff	supervision of work and administration
Inspectors	routine inspection
Overseers	in charge of men and machines
Machine operators Drivers Tradesmen – mechanics – welders – carpenters Labourers Watchmen	routine and emergency work

13

HEALTH AND IRRIGATION

13.1 Introduction
Irrigation can bring wealth and prosperity to a community but it can also bring sickness and sometimes death. Many of the most important diseases in the tropics are related to water in some way. When a new dam or an irrigation scheme is built it can lead to an increase in the prevalance of diseases such as malaria, schistosomiasis. New villages and towns built around the new schemes may increase urban diseases such as filariasis and diarrhoea. There are more than 37 such diseases linked with water. Young chidren often die if there is no medical care. Older people may suffer greatly and are unable to work and live their normal lives.

In many countries not enough attention is being given to these health problems and so as more irrigation schemes are constructed more people may become sick. Those involved in irrigation might think that diseases are a problem only for the health workers. However, there are ways in which irrigators can help to reduce the risks of infection by the irrigation practices they use and the way in which they operate and maintain their irrigation systems.

13.1 Countries where schistosomiasis is a problem.

Those in irrigation can help to fight disease but to fight properly it is important to know something about the enemy. The two most important enemies are Schistosomiasis and Malaria. Each of these diseases has a special life cycle which involves water.

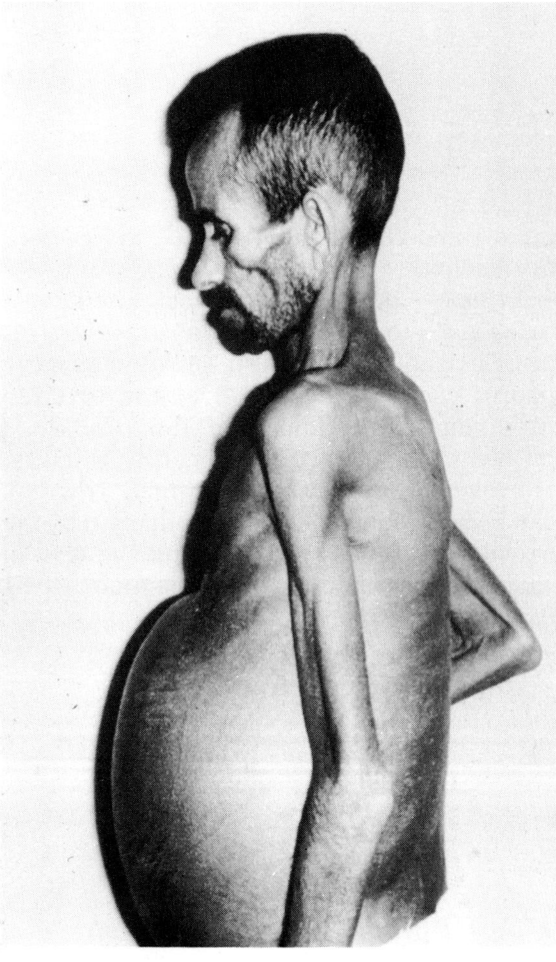

13.2 Man suffering from acute schistosomiasis.

13.2 Schistosomiasis

This is sometimes called *Bilharzia* and is found in many countries (figure 13.1). It is estimated that more than 200 million people are suffering from this disease.

In its early stages victims sometimes have a skin rash shortly after infection and two or three months later have diarrhoea with blood in their stools and urine. They feel very tired and are unable to work properly. Eventually the disease destroys vital organs in their bodies and this results in death. This does not usually happen suddenly. It can take many years (figure 13.2).

The disease is caused by small parasitic worms which live in the human blood stream near the bowel and bladder of an infected person. They lay eggs and some find their way into the bowel and bladder and leave the body in stools and urine (figure 13.3). In places where sanitation is poor, people tend to use canals and drains for their toilet, particularly if they are full of reeds and weeds in which they can be private. The eggs hatch in the water and the free swimming *miracidium* look for suitable fresh water snails in which they can live and develop. The snails are called *hosts* as they house the miracidum like a hotel for 4-8 weeks. Sometimes they are called *vectors* because they help to transmit the disease. Each miracidium can produce thousands of larvae called *cercarae.* These are very small (0.5 mm long) and swim around in the water (figure 13.3). They can easily penetrate the skin of any person who happens to be standing in the water. They find their way through the blood stream to the bowel and bladder and the whole cycle starts again.

13.3 Malaria

This is perhaps the most well known of all the water related diseases in the tropics and affects more than 500 million in the world (figure 13.4). It is a disease that produces high fevers from which many young children die. Those who survive can have continuing attacks. These tend to leave people weak and open to other kinds of infection.

The disease is linked to water because it is carried by *Anopheles* mosquitos. These like to breed in pools of standing water. Undrained fields, badly maintained canals and drains and borrow pits filled with water provide ideal breeding places.

A mosquito picks up the infection when seeking a blood meal and bites a person who is already suffering from malaria. It passes on the disease to an infected person when it takes another blood meal (figure 13.5).

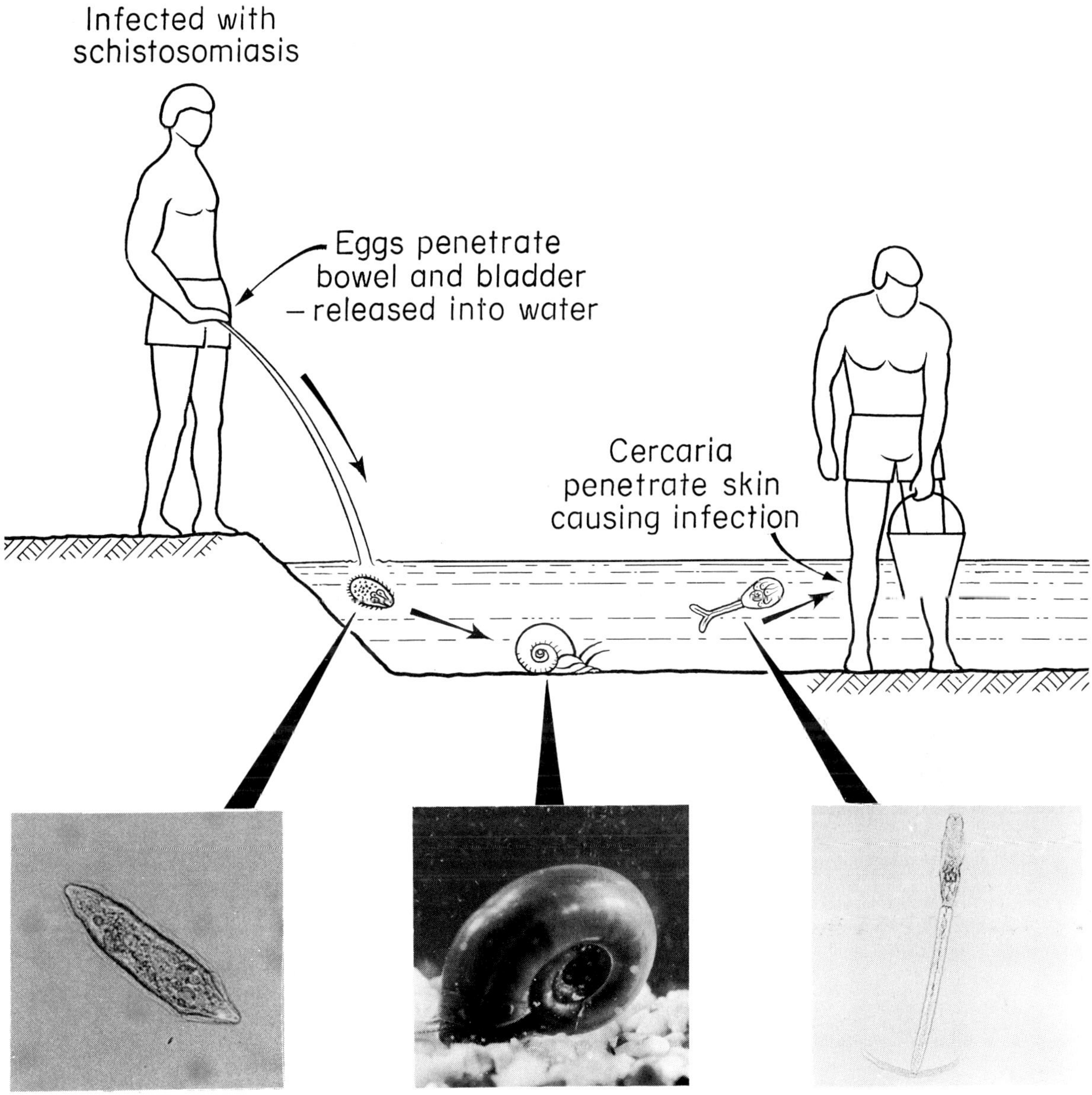

(a) Miracidium hatched from eggs

(b) Snail releasing cercarae (host or vector)

(c) Cercaria (0.5 mm long)

13.3 Life cycle of schistosomiasis.

13.4 *Countries where malaria is a problem.*

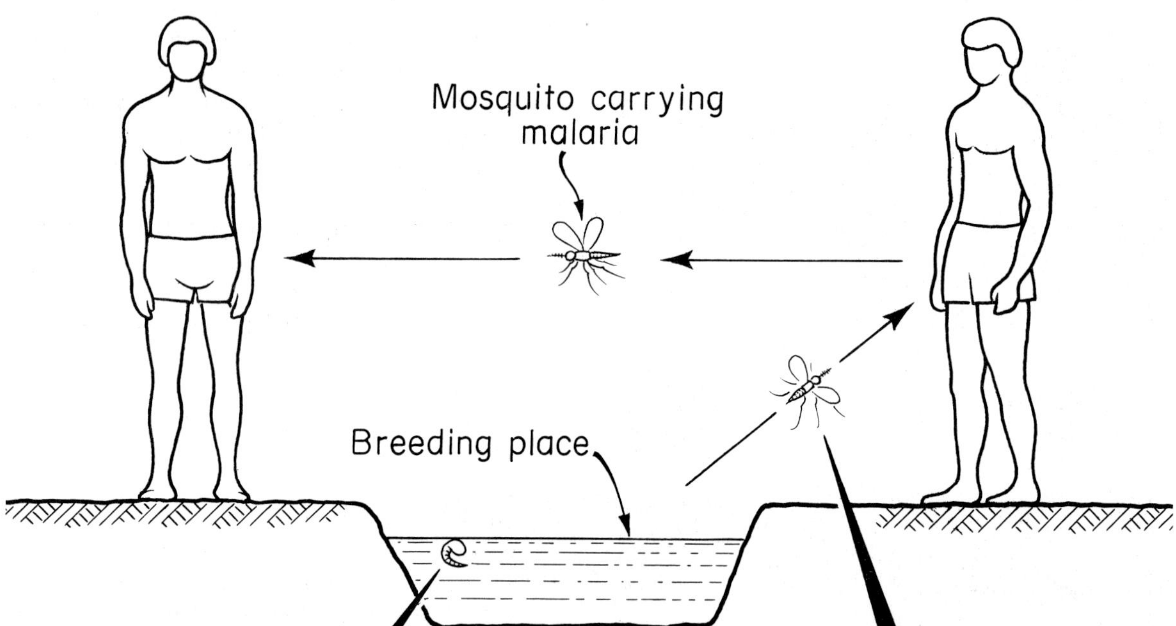

(a) Mosquito pupa in water

13.5 *Life cycle of malaria.*

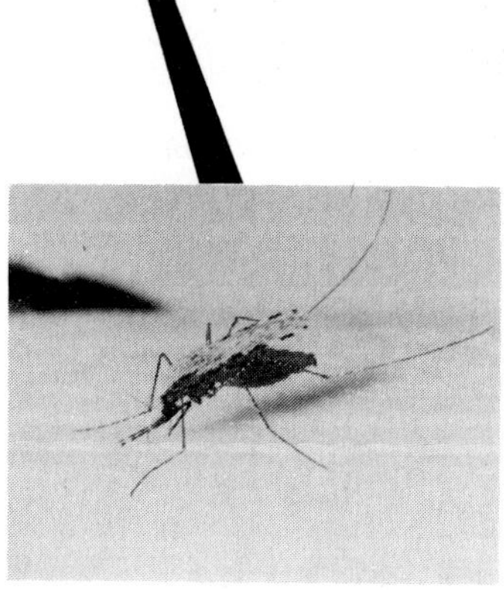

(b) Anopheles mosquito biting

METHODS OF CONTROL

The best way to fight these diseases is to try and interrupt their life cycle. If one part of the cycle is destroyed the disease cannot survive. There are three ways of doing this:

Treat infected people medically

Get rid of the vectors (snails or mosquitos).

Separate people from infested water.

13.4 Medical treatment

If people can be cured of diseases they cannot pass on any infection to others. Drugs are available to cure schistosomiasis and malaria. However, they are very expensive and not always available in the rural areas in the tropics.

13.6 Dripping molluscicide into irrigation water to kill snails

13.5 Getting rid of the vectors

This is where those involved in irrigation can help. If snails and mosquitos are killed the diseases they carry cannot survive.

Snails

Snails like to live in shallow, still or slow moving water in canals, drains or irrigated fields. They prefer channels rich in silt, organic matter and vegetation such as occur on many irrigation schemes (figure 12.2).

One method of killing the snails is to use chemicals called *molluscicides* in the irrigation water (figure 13.6). However, like the drugs they are very expensive and difficult to obtain. Another method is to change the conditions in the channels and fields so that the snails do not like them. They then stop breeding and either move away or die. There are many ways of doing this:

Increase the velocity in channels. Snails like slow moving water. If the velocity is increased to 0.5 m/s the snails are much less likely to survive.
Line canals. This is a very effective method. Water velocity can be much higher, seepage, silting and vegetation growth are reduced. Lining can be very expensive but it may be possible to line canals close to villages when only limited money is available.
Remove vegetation from channels. Snails like to breed in the weeds and they are a source of food.
Clean and straighten channels. Meandering channels can create small stagnant pools.
Backfill or drain borrow areas. These are often neglected places filled with stagnant water.
Inspect canal banks for seepage.
Inspect canals and structures for still-water places where snails can breed.
Check for wasted water (see chapter 11). Excess water may cause flooding and provide breeding sites.
Grade out low-lying land in fields (see chapter 7). This will also improve irrigation efficiency.
Drain-down channels and let them dry out. The dry conditions and the heat from the sun can kill the snails.

Mosquitos

Mosquitos like to live in very similar conditions to snails. However they prefer to

13.7 Providing a good alternative water supply can reduce disease

come in the house at night and tend to rest on the walls after a blood meal.

One method of killing mosquitos is to spray house walls with DDT. Some mosquitos are now resistant to this and so Malathion is used. This is regarded as a very safe insecticide. Like the molluscicides these chemicals are expensive and not always available when they are needed.

Mosquitos can also be killed by changing their living conditions so that they can no longer breed. Many of the methods listed for getting rid of snails will also remove mosquitos.

13.6 Separating people from water

If people do not come in contact with infested water they cannot be infected with schistosomiasis. If they are some distance from water the mosquitos cannot reach them and so they may not get malaria.

There are many ways of doing this:

Ensure that all houses are at least 500 m from the nearest canal or drain;

Erect fencing along channels near villages.

Use pipes instead of channels on small canals near villages.

Provide good foot bridges across channels.

Use sprinkler or trickle irrigation which uses piped water.

In places where people rely on the canals for drinking, cooking and washing water (figure 13.7) it is **essential** to provide a good alternative water supply. If not they will break the fences and pipes to obtain water. Small swimming pools may also be needed in the villages as canals seem to be good places to swim on a hot day!

13.7 Combined approach

Fighting disease is a very difficult task. Each of the methods described will help but they are of limited value on their own. The best results can only be obtained by using all the methods available at the same time.

INDEX

Figures in italics refer to illustration pages